Energetics of Secretion Responses

Volume I

Editor

Jan-Willem N. Akkerman, Ph.D.
Director
Laboratory of Haematology
University Hospital Utrecht
Utrecht, The Netherlands

CRC Press
Taylor & Francis Group
Boca Raton London New York

CRC Press is an imprint of the
Taylor & Francis Group, an **informa** business

First published 1988 by CRC Press
Taylor & Francis Group
6000 Broken Sound Parkway NW, Suite
300 Boca Raton, FL 33487-2742

Reissued 2018 by CRC Press

© 1988 by Taylor & Francis
CRC Press is an imprint of Taylor & Francis Group, an Informa business

No claim to original U.S. Government works

A Library of Congress record exists under LC control number: 88000278

Publisher's Note
The publisher has gone to great lengths to ensure the quality of this reprint but points out that some imperfections in the original copies may be apparent.

Disclaimer
The publisher has made every effort to trace copyright holders and welcomes correspondence from those they have been unable to contact.

ISBN 13: 978-1-138-50614-5 (hbk)
ISBN 13: 978-1-138-55868-7 (pbk)
ISBN 13: 978-1-315-15097-0 (ebk)

Visit the Taylor & Francis Web site at http://www.taylorandfrancis.com and the CRC Press Web site at http://www.crcpress.com

THE EDITOR

Jan-Willem N. Akkerman, Ph.D. is the Director of the Laboratory of Haematology in the Department of Haematology at the University of Utrecht, The Netherlands. He received his Ph.D. degree in 1975 on a thesis entitled "Human Platelet Glycolysis" and was a visiting staff member at the Thrombosis Research Center at Temple University, Philadelphia,Pa. in 1978.

His research efforts are in the field of thrombosis and atherosclerosis, with special emphasis on the role of blood platelets and the mechanisms that make these cells responsive to platelet-activating agents. His current research interests are the energetics of platelet functions, mechanisms for signal transduction, and abnormal platelet behavior in patients with bleeding disorders or hyperaggregability.

CONTRIBUTORS

Volume I

Jan-Willem N. Akkerman, Ph.D.
Director
Laboratory of Haematology
University Hospital Utrecht
Utrecht, The Netherlands

Lilly Y. W. Bourguignon, Ph.D.
Professor
Department of Anatomy and Cell Biology
School of Medicine
University of Miami
Miami, Florida

Genevieve Herman, Dr.Sc.
Charge de Recherche CNRS
Laboratoire de Biochimie
Université Paris-Sud
Orsay, Cedex, France

Holm Holmsen, Ph.D.
Professor
Department of Biochemistry
University of Bergen
Bergen, Norway

Philippe Mauduit, Dr.
Charge de Recherche CNRS
Laboratoire de Biochimie
Université Paris-Sud
Orsay, Cedex, France

Bernard Rossignol, Dr.Sc.
Professor
Laboratoire de Biochimie
Université Paris-Sud
Orsay, Cedex, France

Shimon Schuldiner, Ph.D.
Associate Professor
Department of Molecular Biology
Hebrew University
Jerusalem, Israel

Elizabeth R. Simons, Ph.D.
Professor
Department of Biochemistry
School of Medicine
Boston University
Boston, Massachusetts

Adrie J. M. Verhoeven, Ph.D.
Postdoctoral Fellow
Department of Biochemistry
Erasmus University
Rotterdam, The Netherlands

Hans V. Westerhoff, Ph.D.
Visiting Scientist
Laboratory of Molecular Biology
National Institute of Diabetes, and
 Digestive and Kidney Diseases
National Institutes of Health
Bethesda, Maryland

TABLE OF CONTENTS

Volume I

TABLE OF CONTENTS

Volume II

Energy-Dependent-Steps in Signal Processing

Energy-Requirement of Exocytosis

Energy Metabolism During Secretion

Chapter 1

INTRODUCTION

Jan-Willem N. Akkerman

The last few years have seen a progressive increase in our understanding of the molecular processes underlying secretion responses. The realization that cellular communication is essential for virtually all life forms and that this communication is mediated via chemicals secreted from specific cells has focused attention to the regulation of these responses. With the identification of the many reactions that are initiated when a secretagogue makes contact with its receptor came the understanding that metabolic energy plays a crucial role in the extrusion of granule components.[1-3]

A parallel and equally rapid progress has been made in our knowledge of cellular energy metabolism. Starting from the early work on ATP resynthesizing pathways and the identification of flux-controlling mechanisms, the concepts of phosphate potential and adenylate energy charge have emerged as central dominators in the regulation of supply and demand of metabolic energy.[4,5]

This volume describes the energetic requirements of the various steps in agonist-induced secretion processes. Following a general introduction of the thermodynamic aspects of energy producing and consuming sequences in the cell, the synthesis and storage of secretable products are discussed which form part of the many requirements that must be met before the cell can respond adequately to a secretagogue. Then, the rapidly expanding field of stimulus-secretion coupling is presented, starting with the initial contact between an agonist and its receptor, followed by the generation of second messengers, finally leading to the mechanisms that execute exocytosis of granule material. The process of liberating granule material into the extracellular space is presented separately paying special attention to chemiosmotic processes, membrane fusion, and contractile mechanisms. Finally, the alterations in energy producing and consuming pathways and changes in the concentration of ATP are discussed which form the basis for a technique for the quantitative assessment of the energy requirements of secretion responses.

REFERENCES

1. **Conn, P. M., Ed.,** *Cellular Regulation of Secretion and Release,* Academic Press, New York, 1982.
2. **Case, R. M., Lingard, J. M., and Young, J. A.,** *Secretion, Mechanism and Control,* Manchester University Press, Manchester, England, 1984.
3. **Berridge, M. J.,** The molecular basis of communication within the cell, *Sci. Am.,* 253, 124, 1985.
4. **Atkinson, D. E.,** *Cellular Energy Metabolism and its Regulation,* Academic Press, New York, 1977.
5. **Reich, J. G. and Selkov, E. E.,** *Energy Metabolism of the Cell. A Theoretical Treatise,* Academic Press, New York, 1981.

Chapter 2

CELLULAR ENERGETICS

Hans V. Westerhoff

TABLE OF CONTENTS

I. INTRODUCTION

Living organisms are presented with numerous challenges by the physical and chemical principles of this world. For one of these, i.e., that most organic chemical reactions solely proceed at appreciable rates at temperatures far above the boiling point of water, scientists have found the solution of creating highly effective catalysts. For a second, i.e., to remember how they were built such that their progeny will be built according to the same, well-tested principles, the deoxyribonucleic acids have been invoked. With respect to a third, i.e., to keep unwelcome substances outside and precious substances inside and concentrated, evolution brought biological membranes. Yet, there remain a number of boundary conditions which cannot be solved by simple tricks. One of these is the problem that, in principle, "life" is somewhat contrary to the natural tendency of physical chemical processes. Whereas the latter tend to increase randomness and decrease the energy content, a growing organism tends to increase in ordering and in energy content.

Processes that run against the direction physical chemistry might predict, are rather common although not unique to living cells. The obvious examples are the biosynthetic processes, the synthesis of ATP, and the transport of potassium into and sodium out of the cell. The reader of this book will be interested in the secretion of substances from cells. In this case energy is involved in the storage of substances to be secreted into secretory vesicles, but also included in this book[1,2,3] is the secretion itself. Before reviewing aspects of the various secretory processes, it is perhaps useful to review in this chapter some of the general principles of biological energetics. Although some of these general principles can be found in general textbooks on biochemistry and bioenergetics,[4-7] there are a number of recent developments. We shall include the latter.

We shall begin with reviewing the differences between energy, enthalpy, entropy, and free energy and why the latter is the one that is most informative for our purposes. Next, the different forms of free energy encountered in biological systems and the most important interconversion reactions will be discussed in terms of a free-energy map of the living cell. To allow for sufficient efficiency of these free-energy conversions, processes of sometimes quite different natures must be coupled reasonably well. Coupling of the reactions catalyzed by disparate enzymes through a common metabolite, as well as coupling within a single enzyme will be discussed. The various methods to assay the "energy state" the cell is in will be compared.

In addition to driving reactions in the (necessary) direction in which they by themselves would not proceed, cellular free energies can also play the roles of catalysts and signals. We shall discuss observed correlations between cellular activities and energy state parameters in terms of their possible origin. We shall end with some speculations of the involvement of membrane potentials and osmotic pressures in fusion and secretion.

II. THE VARIOUS FORMS OF ENERGY

A. Energy and the First Law

There are a number of strict principles chemical and physical processes must adhere to.

A simple one is that of conservation of the elements. In anaerobic bacterial growth for instance, the conservation of the element carbon, together with the rather limited range of alternative products some cells can excrete, almost completely determines the growth yield.[8,9] In more complex chemical reaction systems such as the eukaryotic liver cell, the number of possible reactions is so large that this same restriction is hardly felt. Under some conditions however, only a limited number of reactions occur at a significant rate, with the effect that moieties, such as the adenosine-monophosphate group, are conserved; if only ATPases, kinases, adenylate kinases, and phosphatases (not of AMP) are active, the sum concentration of ATP, ADP, and AMP is conserved.[10] (see also References 3 and 11 and below). As biochemists, however, we have grown accustomed to doing the chemistry "right", i.e., such that in all proposed metabolic pathways the chemical conservation conditions are obeyed.

There is another property however, that is also conserved, but about which we hardly worry. This is the energy, denoted by the symbol U. The law that says that energy can neither be produced, nor consumed, but only transformed or transported, is, of course, the first law of thermodynamics.[7] If biological systems were closed with respect to heat exchange with their environment, we would have to worry about this law. However, in by far the most cases, we consider systems that are "thermostated" by a heat bath, which freely allows the exchange of heat between the biological system and its environment until the two are at the same temperature. If the biological system would convert a compound of high energy to a compound of low energy, then the liberated energy would not have to remain within the system and heat it up until it would denature, but the excess energy would be readily excreted to the environment. If, more rarely, the biological process would be endothermic, the system would not have to cool down with the effect that its reactions would slow down; heat flow from the environment would make up for the decrease in energy.

B. Free Energy and the Second Law

Why then bother about thermodynamics in biological systems? Well, there is a principle, called the second law of thermodynamics, for which there is no such simple reservoir solution. Also, we do not automatically take care of the corresponding restriction, even if we obey all our rules of chemistry in terms of element or moiety conservation. For instance, if we would have identified in our cell free extract, the enzymes lactate dehydrogenase, enolase, phosphoglyceromutase, phosphoglycerate kinase, glyceraldehyde 3-phosphate dehydrogenase, aldolase, triose phosphate isomerase, phosphofructokinase, phosphoglucose isomerase, and hexokinase, we might postulate the existence of a phosphoenolpyruvate phosphatase which would then complete the pathway for the synthesis of glucose from two molecules of lactic acid. The second law of thermodynamics would prevent our peers from funding this research proposal, because the flux in such a pathway would not be in the desired direction. Or, closer to the interest of most readers of this book, a proposal of a carrier of singly protonated catecholamine, as the basis of the accumulation of this substance in chromaffin granules, though otherwise chemically and physically sound, would not stand the test of thermodynamics.

The principle embodied by the second law of thermodynamics is that not all energy can be used to do work. Perhaps the simplest example is the law that, whenever we wish to do work, we cannot simply extract heat from our environment (which is supposedly at the same temperature) and use that as the energy source for the work.

Thermodynamics has been a significant help to biologists by defining a new type of energy (the free energy) that solely consists of the energy that can be applied to do work. I will now review how this free energy "corrects for" the energy that cannot be used to do work.

For biological systems it is essential that they are open with respect to the exchange of heat, volume, and chemical substances with their environment. The incremental change in system energy, dU, may be formulated in terms of these exchanges as follows:[7]

$$d_eU = T \cdot d_eS - P \cdot dV - f \cdot dl + \mu_j \cdot d_en_j \qquad (1)$$

Here the subscripts e stress that we are considering the component due to exchange only. For the energy, this makes no difference; since it is a conserved property, d_eU always equals dU. The first term on the right in this equation represents the exchange of heat because of the definition of entropy:

$$d_eS = dQ_{reversible}/T \qquad (2)$$

(Heat exchange is taken to proceed reversibly. This is an unproblematic assumption for biological systems.[7]) The second term in Equation 1 describes the volume work, which is usually quite useless. The second last term describes the increment in system energy due to work done by a contractile element of the system (muscle), with force f and length l. The last term describes the increase in energy due to the import of an amount d_en_j of substance j, at chemical potential μ_j.

In the usual system with uniform and constant hydrostatic pressure and uniform and constant temperature, the amount of useful work that can be done is given not by the energy U, but by the Gibbs free energy, G.[12] We owe this to the definition of the Gibbs free energy:

$$G = U + P \cdot V - T \cdot S \qquad (3)$$

where P, V, T, and S refer to the pressure, volume, absolute temperature, and entropy of the system, respectively. For the change in free energy is given by:

$$dG = dU + P \cdot dV + V \cdot dP - T \cdot dS - S \cdot dT \qquad (4)$$

which combines with Equation 1 as:

$$dG = V \cdot dP - S \cdot dT - T \cdot d_iS - f \cdot dl + \mu_j \cdot d_en_j \qquad (5)$$

The terms corresponding to heat exchange ($T \cdot d_eS$) and the unuseful volume work ($P \cdot dV$) have dropped out and only the useful work terms referring to work of contraction and transport have been retained. At constant temperature and pressure the first two terms on the right-hand side of Equation 5 disappear.

The subscript i in $T \cdot d_iS$ refers to the change due to production of the property inside the system. To derive Equation 5 it was used so that the total increase in a property of the system (d) would equal the increase due to production (d_i) plus the increase due to net import (d_e). The second law of thermodynamics implies that, contrary to energy U, entropy and free energy are not conserved properties. Entropy can be produced though never destroyed. Consequently, the term $-T \cdot d_iS$ is always smaller or (but only if the system is at equilibrium, which is unrealistic for biological systems) equal to zero. It is this term that corresponds to the intuitive notion that spontaneous processes always go downhill in terms of "energy" (or that they must "liberate energy"); inside the system an amount of free energy $-d_iG = T \cdot d_iS$ must be demolished whenever processes occur.

Recognizing the latter two terms of Equation 5 as terms corresponding to the exchange of free energy with the environment and assuming that pressure and temperature are constant, we may rewrite this equation as:

$$-dG + d_iG = -d_eG = +f \cdot dl - \mu_j \cdot d_en_j \qquad (6)$$

$-d_iG$ is the term corresponding to the free-energy dissipation inherent in ongoing processes which is always positive (or zero, but only at equilibrium). Considering the term f · dl as representative of the work the biological system can do, we see that the amount of work that is done by the system is given by the decrease in its free energy plus the free energy that is gained by the import of a substance with a high chemical potential, minus the amount of free energy dissipated in the process.

In fact, Equation 6 describes the energetics of living systems (or parts thereof) reasonably well. First, living systems tend to grow, which often means that they increase in ordering (i.e., decrease in S) and increase in energy content. Both these increases lead to an increase in the free energy of the organism (i.e., −dG < 0, cf. Equation 3). Second, these organisms do mechanical work, such as in locomotion and muscle contraction. This implies a positive f · dl term. Third, living systems and parts thereof produce and secrete valuable substances, i.e., there is a negative $d_e n_j$ for substances with a high μ_j. Also, this contributes a positive term to the right-hand side of Equation 6. Finally (or, not quite, see below), the fact that processes are at all occurring in the living system, implies that free energy must be dissipated, i.e., that $d_i G$ is negative. The energetic dilemma for living systems is therefore that it is to export free energy (negative $d_e n_j$ for compounds j with high chemical potential), i.e., that the right-hand side of Equation 6 must contain positive terms, whereas its free-energy content should increase (or remain constant) and it should also destroy some free energy, such that the left-hand side of Equation 6 tends to be negative.

There are two ways in which the organism may attempt to meet this problem. First, it may use an internal store of free energy in terms of metabolizable substances, which it inherited from its mother. By decreasing the free energy of this store, − dG and hence the left-hand side of Equation 6 can indeed become positive. In the long run however, stores cannot be sufficient. The long term solution must therefore be that the organism imports substances of high chemical potential and exports "products" of low chemical potential. This will produce a negative term in the right-hand side of Equation 6, thus allowing the organisms to meet the requirements imposed by this equation.

C. Enthalpy

The enthalpy is a less useful property of a system than the free energy, although it is often much more accessible. Defined as U + P · V it only corrects for the useless volume work but not for the fraction of energy that cannot be used to do work because it is pure heat. For the increment of enthalpy of a system we find

$$dH = V \cdot dP + T \cdot d_e S - fdl + \mu_j \cdot d_e n_j \tag{7}$$

This equation shows that at constant pressure and in the absence of contraction work and transport into and out of the system, the change in enthalpy is equal to the amount of heat exchanged with the environment. Consequently, it is easily measured by calorimetry. However, since

$$dG = dH - T \cdot dS \tag{8}$$

enthalpy changes are only indirectly related to the more relevant (see below) changes in free energy.

III. DIFFERENT FORMS OF FREE ENERGY THAT ARE RELEVANT FOR BIOLOGICAL SYSTEMS

A. The Primary Free-Energy Sources for Cells

From mechanisms we know that the capacity to do work can reside in many different

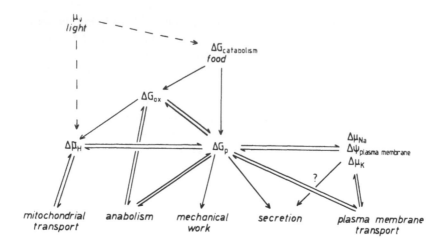

FIGURE 1.   A map of cellular free-energy metabolism. ΔG_{ox} refers to the free energy of oxidation of redox coenzymes like NAD(H), $\Delta\bar{\mu}_H$ to the electrochemical potential difference for protons across the inner mitochondrial membrane and ΔG_p to the phosphate potential, or free energy of hydrolysis of ATP. Anabolism comprises all biosynthetic chemistry.

configurations. One is that of a compressed spring. Another is that of a leaden ball on a roof. In biology, there are three major forms in which free energy can appear. The purest form, in the sense that it completely disappears once its free energy has been dissipated, is the photon. However, as a direct source of free energy, photons are not important for animals. This leaves us with the other two: free energy residing in a chemical reaction and free energy contained in an (electrochemical) concentration gradient across a membrane.

As shown by Equation 6, a system can obtain some of the free energy it needs by importing substrates (S) of high chemical potential and excreting products (P) of lower chemical potential. To make this possible, the system will have to convert the former to the latter in a chemical reaction. The free energy imported is equal to:

$$\Delta G_c = \mu_S - \mu_P \tag{9}$$

which is equal to the free-energy difference of the chemical reaction. The conversions that play this role for the entire cell are often grouped together as "catabolism".[13] The free-energy differences across the catabolic reactions have been indicated by ΔG_c, the index c standing for catabolic. Examples are the conversion of glucose to lactate or to CO_2 and the conversion of fatty acids to CO_2. In almost all nonphotosynthetic organisms this is the sole type of free-energy source for the cell as a whole. The free energy "harvested" from the catabolic reactions is, however, rarely directly applied to biosynthetic processes or "work". It is first converted to intermediary stages as illustrated by Figure 1.

B. Intermediary Free-Energy Differences

Many catabolic processes contain oxidation reactions that do not use molecular oxygen as the oxidant but a redox coenzyme, mostly NAD^+ (or $NADP^+$ or a flavoprotein). Thus, part of the free energy of the oxidation reaction is conserved in the free energy of the reaction:

$$NADH_2^+ + \tfrac{1}{2}O_2 \rightleftharpoons NAD^+ + H_2O \tag{10}$$

This free-energy difference is often called ΔG_{ox}. Part of it is funnelled directly into biosynthesis at steps where a substrate must be reduced.

Much of the free energy temporarily stored in the redox free-energy difference is however converted to another free-energy difference. Here we encounter the third type of free energy relevant for biological systems, i.e., that contained in gradients across membranes; the one relevant here is the electrochemical potential difference for protons across the inner mitochondrial membrane, $\Delta\bar{\mu}_H$, consisting of the pH gradient across that membrane, ΔpH, and the electric potential difference across that membrane, $\Delta\Psi$:

$$\Delta\bar{\mu}_H = 5.7\Delta\, pH + F\Delta\psi \tag{11}$$

This conversion is catalyzed by the electron transfer chain located in the inner mitochondrial membrane. An enzyme located in the same membrane, i.e., the $\overrightarrow{H}^+$-ATPase, converts part of the free energy contained in $\Delta\bar{\mu}_H$ into the even more universal free-energy currency for the cell: the free energy of hydrolysis of ATP, ΔG_P (also called the "phosphate potential"). The $\overrightarrow{Na}^+,\overleftarrow{K}^+$-ATPase of the plasma membrane of eukaryotic cells converts part of this free energy to an electric potential difference and concentration gradients of both Na^+ and K^+ across that membrane.

C. Output Free-Energy Differences

As indicated in Figure 1, both $\Delta\bar{\mu}_H$ and ΔG_P serve as free-energy donors for biosynthesis. The role of the former is centered somewhat more around transport steps across the inner mitochondrial membrane and the role of the latter is more centered in energizing thermodynamically uphill chemical transitions and mechanical work. Although for some time a major role of $\Delta\bar{\mu}_H$ in secretion was considered[14] through driving the swelling of secretory vesicles and the accumulation of anions in them, a more direct role of ATP is now envisaged.[1,15] The free energy present in ATP can also be partly converted to free-energy of hydrolysis of creatine phosphate. In this form, and, in bacteria in the form of a transmembrane sodium gradient, free energy is stored by the cell for intermediately long times.

IV. FREE-ENERGY CONVERSION AND COUPLING

A. Why Coupling is Important

In Equation 6 we stressed that cells would not be able to do what they need to do if they would not compensate for the free energy cost by taking up high free-energy substrates from the environment and secreting low free-energy products. Viewing the cell as a free-energy converter, we consider the latter conversion of substrates to products as the input reaction. Let us set out by considering this input reaction as a process that proceeds completely independent of other processes. In converting the substrate for the catabolic "input" reaction, S_{in}, to the product of this reaction, P_{in}, at a rate J_{in}, the free energy the system absorbs from the environment is $d_e G_{in}/dt = J_{in} \cdot \Delta G_{in}$. If a system is in steady state, its properties including its free energy do not change between subsequent turnovers. From Equation 6 we find that the rate of free energy input must equal the rate at which the free energy is dissipated, i.e., all input free energy would simply be destroyed.

The biological system usually has to carry out processes that are uphill in terms of free energy. We shall call these processes "output processes". If we also consider every output process as completely independent from any other process, then it donates the following free-energy to the environment: $-d_e G_{out}/dt = -J_{out} \cdot \Delta G_{out}$, where ΔG_{out} is positive and J_{out} is negative in the desired direction of actual output. If the output process is in steady state this must be equal to the free-energy dissipation by the output process and, hence, by virtue of the second law of thermodynamics, be positive.

It follows that the second law of thermodynamics would forbid the output process to occur

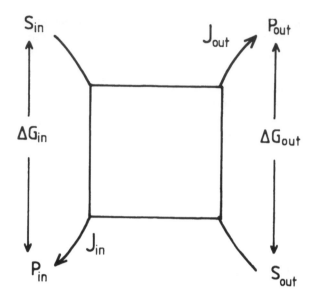

FIGURE 2.   Diagram for a free-energy converter. The converter catalyzes a thermodynamically downhill input reaction (such as catabolism, respiration) and a thermodynamically uphill output reaction (such as cellular growth, ATP synthesis). The latter is possible because of the transduction of free energy from the input reaction to the output reaction.

in the desired direction (i.e., with negative J_{out}). It is essential for biological systems that they deviate from the starting point of the above arguments, i.e., from the assumption that the input and the output process are independent of each other.[16] For biological systems coupling between input and output processes is essential: it is necessary that some of the free energy "liberated" in the catabolic "input" reaction is actually transduced to the anabolic "output" reaction. In fact, this is true for all the steps in Figure 1 where free energy is transduced from one form to another. Figure 2 gives a general scheme for each free energy transducing process.

B. Characteristics of Coupling

For the case that the free-energy transducer (the square in Figure 2) is in steady state, we find:

$$- d_i G/dt = J_{in} \cdot \Delta G_{in} + J_{out} \cdot \Delta G_{out} \tag{12}$$

where we now accept that the right-hand term can be negative, provided that the left-hand term is more positive and that this can occur only because there is some mechanism that couples the input and output reactions.

For the case of a free-energy transducer that is near equilibrium, the coupling can be described in a simple way:[7,17]

$$J_i = L_{ii} \cdot \Delta G_i + L_{io} \cdot \Delta G_o \tag{13}$$

$$J_o = L_{oi} \cdot \Delta G_i + L_{oo} \cdot \Delta G_o \tag{14}$$

Here the L_{jk} are proportionality coefficients which are independent of the free energies of reaction. L_{oo} and L_{ii} are positive. Owing to our sign convention, either free-energy difference

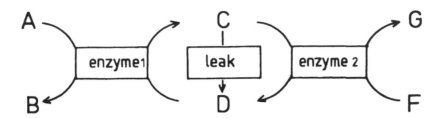

FIGURE 3. Coupling of reactions catalyzed by separate enzymes 1 and 2. The leak, by causing just the conversion from C to D, causes uncoupling.

is positive in free-energy transducing systems. Hence the output flow can only become negative (which means that it would be in the desired direction) if L_{oi} is sufficiently negative. For reasons of normalization the negative, coupling coefficient q, has been defined as $L_{oi}/\sqrt{(L_{oo}/L_{ii})}$. Onsager's reciprocal relations imply that $L_{oi} = L_{io}$.[7,18] Consequently, the fact that L_{oi} is negative immediately implies that the input reaction responds to changes in the free-energy difference across the output reaction. By use of the definition of the coupling coefficient q and the phenomenological stoichiometry $Z (= \sqrt{(L_{oo}/L_{ii})})$, the above equations lead to the following expression for the ratio of the output flow to the input flow:

$$j = -J_o/J_i = Z \cdot (q + Z \cdot \Delta G_o/\Delta G_i)/(1 + q \cdot Z \cdot \Delta G_o/\Delta G_i) \qquad (15)$$

This equation shows that, if there is coupling ($-1 \le q < 0$), there should be an output flow induced whenever an input flow is induced. However, the ratio between the two flows may vary depending on the free energies of the two reactions to be coupled, unless coupling is complete ($q = -1$).

Thus, we may conclude that, near equilibrium, coupling has a number of implications: first, the output reaction can run against its own free-energy difference. Second, the input reaction rate is sensitive to changes in the output driving force and vice versa. This may be called the principle of force coupling. Third, if one induces the input reaction without touching the degree of coupling, an output reaction should occur, and vice versa. This may be called the principle of flux coupling. Strictly speaking, this analysis is only straightforward when the system is close to equilibrium. For the farther from equilibrium case (which is the one that corresponds to most biological processes), the second of these implications may not be necessary; the processes may be saturated with respect to one of the two free-energy differences, but still be coupled.[7,13] The extension of nonequilibrium thermodynamics to the actual biological processes that tend to be farther from equilibrium has been reviewed in Reference 7.

C. Coupling at the Two-Enzyme Level

Two chemical reactions may be coupled because they share a metabolite. In the example of Figure 3 enzyme 1 while converting A to B (e.g. glucose to glucose-6-phosphate) also forms C from D (e.g., ADP from ATP). Enzyme 2 forms G from F while converting C back to D. With ADP and ATP taking the role of C and D respectively, this is indeed one of the most important coupling mechanisms in cellular metabolism. It couples biosynthetic pathways to catabolism (see also Figure 1). The free-energy transduction is rather clear because there is a readily definable free energy intermediate, i.e., the chemical potential difference between C and D.

With C and D representing protons on the inner and outer side of the inner mitochondrial membrane, respectively, Figure 3 would represent the chemiosmotic coupling mechanism of oxidative phosphorylation.[6] A would then correspond to $NADH_2^+$ plus $^1/_2O_2$. B would

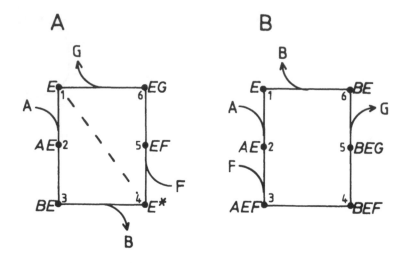

FIGURE 4.   Coupling by a single enzyme. The enzyme couples the conversion of
A to B to the thermodynamically uphill conversion of F to G. Enzyme states[20] are
indicated in italics and numbered. (A) and (B) give different mechanisms.

correspond to NAD^+ plus H_2O, F would correspond to ADP plus phosphate, and G to ATP.
Enzyme 1 would be the electron-transfer chain in the inner mitochondrial membrane and
enzyme 2 the $\overrightarrow{H}^+$-ATPase. In this case it becomes clearer that the coupling between the
reactions catalyzed by enzyme 1 and the reactions catalyzed by enzyme 2 do not have to
be coupled completely; if there is an additional way in which C can be converted to D (by
the "leak" in Figure 3), leakage may occur. In oxidative phosphorylation, this leakage
would consist of passive proton permeability of the inner mitochondrial membrane.

D. Coupling by a Single Enzyme

1. The Kinetics

It may be noted that the conversion of A to B and of D to C, as catalyzed by enzyme 1
in Figure 3 is also an example of a coupled reaction. However, here the coupling is catalyzed
by a single enzyme. Figure 4A gives an enzyme diagram for a sequential ordered mechanism.[19,20] The enzyme proceeds from one state to the next in a sequence of six states. The
differences among states reside in whether ligands are bound and in additional conformational
changes. Notably, states 1 and 4 differ only in conformation. This allows for direct transitions
between these states (the dotted line in Figure 4). Coupling depends on the probability of
these states being relatively small, or on the transition probability between states 4 and 1
being small.[21] In Figure 4A, one may define the difference in chemical potential between
E* and E as the free-energy intermediate in the free-energy transduction by the enzyme.
The coupling would here be purely through a conformational mechanism.[22]

In the mechanism given in Figure 4B, this is no longer possible because the two reactions,
i.e., the conversion of A to B and the conversion of F to G are completely interwoven.
Indeed, Hill[20] has shown that the only essential thing to free-energy transduction by a fully
coupled enzyme cycle like that given by the full lines in Figure 4, is that the free energy
along the cycle decreases. It is not essential that the free-energy changes are confined to
particular sites in the cycle.

A slightly more involved case than that of Figure 4A is given in Figure 5, which has one
additional state (called state 7). Since catalysts have to return to their original state, we only
have to consider the cycles that are possible in the diagram of Figure 5A (cf. Figure 5B).

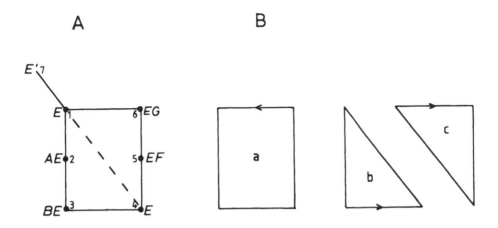

FIGURE 5. Coupling and cycles. (A) An enzyme like that in Figure 4A except that there is an extra state 7 differing from state 1 only in conformation. (B) The three catalytic cycles possible in (A); the arrowhead indicates the direction in which cyclic flux will proceed under the usual conditions.

In cycle a, the conversion of A to B is completely coupled to the conversion of F to G. In cycle b, A is converted to B without concomitant conversion of F to G. Cycle c will tend to operate in the reverse direction converting G to F without regaining A from B. The degree of coupling of the overall reaction will depend on the number of cycles a that occur relative to the number of times b and c are cycled.

For each cycle k, the rate can be found as the mathematical product:

$$J_k/E_{tot} = \Pi_{k+} \cdot (1 - e^{-\Delta G_k/RT}) \cdot \Sigma_k/\Sigma \qquad (16)$$

Here ΔG_k is the free-energy drop along the reaction cycle; for cycle a, it is the free energy of the reaction $A + F \rightleftharpoons B + G$. For cycle b, it would be the chemical potential difference between A and B. Since all other factors in this equation are always positive, it is seen that this free-energy difference determines the direction of the reaction, such that the flux along any cycle always flows downhill in the free-energy sense. This is why ΔG_k is called the thermodynamic driving force of the cycle. It should be noted however, that the magnitude of ΔG_k also affects the absolute magnitude of the cycle flux.

Π_{k+} in Equation 16 is the product of the unidirectional pseudounimolecular (i.e., concentration dependent) forward rate constants. It is a measure of the "catalytic activity" of the enzyme. The term Σ_k/Σ represents the probability that cycle k indeed occurs. This probability would be reduced if, for instance, state 7 would be extremely stable with respect to the other states of the enzyme. It is important to note that depending on how ΔG_k is varied, Π_{k+} and/or Σ_k/Σ may vary with it. Therefore it is difficult to predict for the completely general case how the cycle flux varies with its driving force.[20] It can be shown that if ΔG_k is much smaller than RT, the variation of the factor $1 - e^{-\Delta G_k/RT}$ dominates and becomes linearly proportional.

2. Free Energy as Driving Force, (In)activator, Catalyst, or Cycle Director

For cycle a in Figure 5, the product of the unidirectional forward rate constants can be expressed in first and second order rate constants and the concentrations of A and F:

$$\Pi_{a+} = k_{12} \cdot [A] \cdot k_{23} \cdot k_{34} \cdot k_{45} \cdot [F] \cdot k_{56} \cdot k_{61} \qquad (17)$$

The thermodynamic driving force also depends on the concentrations of A and B:

$$\Delta G_a = [A] \cdot [F]/([B] \cdot [G] \cdot K_{eq}) \qquad (18)$$

It follows that, if we vary the driving force of the reaction catalyzed by cycle a by increasing the concentration of A, the reaction rate will increase, both because the term representing the thermodynamic driving force increases and because the catalytic activity (Π_{k+}) increases. Thus, not only can an increase in a free energy affect reaction rates because it is a driving force, but also because it has a catalytic effect.[23]

We can make this intuitively clear by considering the case where in Figure 5A the transition between states 2 and 3 would be "rate limiting" with respect to the cyclic flux. An increase in ΔG for this cycle could be brought about by either increasing [A], decreasing [B], increasing [F] or decreasing [G]. In this case, increasing [A] would be more effective in increasing the rate than any of the other increases. An increase in [A] here would have a catalytic effect in addition to a driving force effect.

This catalytic effect of the driving force on the reaction rate is somewhat related to the effect of a catalyst, which lowers the activation free energy of a reaction. In the Eyring absolute rate theory, one assumes an equilibrium between the reactants and the transition state complex. As a consequence the reaction rate can be formulated as:[24]

$$v = e^{(-\Delta G^{\#}/RT)} \cdot [S] \cdot k_E \qquad (19)$$

Here $\Delta G^{\#}$ is the standard free-energy difference between the activated state complex and the ground state. k_E is proportional to the absolute temperature and has a value of about $6 \cdot 10^{12}$ sec^{-1} at room temperature. [S] is the concentration of the substrate of the reaction. For the case of Figure 4A, one might equate state 1 to the ground state and state 4 to the activated complex state. In this case the chemical potentials of A and B, i.e., part of the driving force, would enter the term $\Delta G^{\#}$ in the above equation.

There may be cases where state 7 is rather stable compared with state 1. In that case a considerable fraction of the enzyme may be trapped in state 7 and these enzyme molecules will not contribute to the cyclic flux. If however, one would increase ΔG_a by increasing the concentration of A, then the concentration of state 1 would be reduced and therefore (because states 1 and 7 must be at equilibrium) also the concentration of enzymes in state 7 would decrease. In this example an increase in free-energy difference of reaction would direct more enzymes into the catalytic cycle and may therefore have an activating effect.

Finally, changes in the free-energy difference of the total reaction may change the ratio of the fluxes of the different cycles (a, b, and c in Figure 5) with respect to one another. The free energy may serve to direct the enzyme to the proper cycle. Pietrobon and Caplan[25] have analyzed these effects in detail.

V. HOW TRANSMEMBRANE ELECTRIC POTENTIALS MAY AFFECT MEMBRANEOUS ENZYMES

A. Stationary Electric Fields

Because A and B may really stand for ion concentrations at different sides of a membrane, the driving force across the reaction discussed above may contain an electrochemical potential difference for an ion across that membrane, and thus the transmembrane electric potential. As such we could understand that a transmembrane electric potential would influence a reaction rate, either as a driving force, or catalytically, or as an activator, or as a cycle director. Recently, we have come across a rather unsuspected possibility, where a transmembrane electric potential difference could affect the reaction rate of an enzyme that did *not* catalyze the transmembrane movement of an *ion*.[23,26] Figure 6 illustrates the case in point. An enzyme converts the possibly neutral, substances S and P, in a catalytic cycle

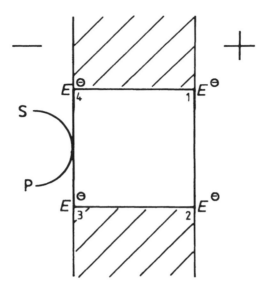

FIGURE 6. Effect of a transmembrane electric field on an enzyme that is embedded in the membrane. It is assumed that in going from state 4 to state 1 or from state 3 to state 2 a negatively charged arm of the enzyme crosses the membrane from left to right. The enzyme is taken to catalyze the interconversion of neutral substances S and P through the cycle 3, 4, 1, 2, 3 etc.[23,26,27]

that involves the cyclic translocation of a charged part of the enzyme across the membrane (or alternatively a cyclic change in dipole moment). It could be shown that a stationary electric field across the membrane could act catalytically on the conversion of S to P.

The reason why such a phenomenon could occur can be understood from an inspection of Figure 6. Suppose that, for some reason, the transition from state 4 to state 1 were "rate limiting" for the cyclic flux. Then the transmembrane electric potential difference of the orientation shown could electrophoretically stimulate the transition from 4 to 1 and hence accelerate the cyclic flux. That this effect of the membrane potential could only be catalytic and could not serve as a force that could drive the conversion of S to P away from equilibrium can be understood by noting that the electric potential would also affect the transition between states 3 and 2. It is only away from equilibrium that the transitions 4 to 1 and 3 to 2 can differ in the extent to which they are limiting the cyclic flux. From these considerations, it can be anticipated that the activities of some transmembrane enzymes are modulated by variations in the membrane potential.[23]

B. Oscillating Electric Fields

The situation would become even more interesting if the transmembrane electric potential difference would oscillate. It has been shown that in such a case, the oscillating component of the membrane potential could serve as a driving force for the catalytic cycle of an enzyme like that in Figure 6.[23,26] Again this can be understood intuitively. In the situation shown in Figure 6, enzymes will tend to go from state 4 to state 1. If state 2 is more stable than state 1, this will be followed by transitions from state 1 to state 2. If then, approximately at the moment that most of the enzymes have reached state 2, the polarity of the field would be reversed, then the enzymes would proceed from state 2 to state 3, and, if state 4 is more stable than state 3, to state 4. Repetition of the procedure would cause clockwise cycling of the enzyme and could thus lead to the production of S from P, even if the chemical potential of S would exceed that of P.

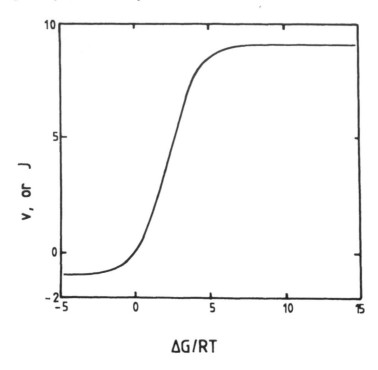

FIGURE 7.   An example of how a reaction rate may vary with the free-energy difference of reaction. The calculation was for the simplest, reversible, enzyme-catalyzed conversion of S to P, under conditions where the sum of [S] and [P] were constant. Details were as in Reference 11, with $V_S = 10$, $V_P = 2$, $K_S = 1$, $K_P = 10$, $[S] + [P] = 10$.

Although the overall electric potential difference across biological membranes may not oscillate much, this may not be so on a local scale; in the vicinity of an electrogenic ion pump, or an ion channel, one would expect the electric field to fluctuate. It has been shown that such field fluctuations could indeed lead to free-energy transduction of this type.[23,27]

VI. DIFFERENT ROLES OF FREE ENERGIES IN THE CELL

Although from the basic thermodynamic point of view, free-energy differences are often looked upon solely as determining in which direction a reaction will flow, we have seen above, that the magnitude of free-energy differences may also affect the absolute magnitude of reaction rates. If the driving force increases, the reaction rate tends to increase. This tendency is a physical necessity near equilibrium, but becomes less inescapable farther away from equilibrium. By way of illustration, Figure 7 gives the relationship for a representative case of enzyme kinetics. For normal enzyme kinetics, reaction rates (1) tend to increase with the free energy of reaction, (2) this increase is only proportional near equilibrium; farther from equilibrium the increase may be proportional, linear, or curvilinear, (3) the relationship between reaction rate and driving force is not unique; it depends on the catalytic activity of the enzyme and on factors such as the total amount of substrate plus product present. It may also be noted that the variation of reaction rate with free-energy difference as calculated from simple enzyme kinetics is reminiscent of kinetic relationships that have been interpreted as being the result of "gating"; apparently, normal enzyme kinetics already leads to the phenomenology of gating, without a special gating mechanism.

In terms of the discussion of the above section, part of the effect of changes in the free-energy difference of reaction could be attributed to driving force effects and parts might be attributed to catalytic effects.

At the level of metabolic pathways there exists a number of rather clear examples of catalytic, or (see above) activation roles of free-energy differences. One noted example is the catalytic effect of the free energy of hydrolysis of ATP on the glycolytic flux. Although the total free-energy balance of glycolysis has $2 \cdot \Delta G_p$ as a counteracting driving force (two ATP being produced per metabolized molecule of glucose), an increase in ΔG_p (at low concentrations of ATP, by increasing the concentration of ATP) can increase the glycolytic flux. This is because part of the ATP hydrolytic free energy is used to bring glucose to the "activated state" of glucose-6-phosphate.

The same pathway illustrates yet another effect of the free-energy difference ΔG_p: at higher magnitudes, an increase of ΔG_p will tend to inactivate glycolysis by allosteric effects on enzymes like phosphofructokinase and hexokinase. Another way in which free energies play a role in activating or inactivating metabolic pathways is the covalent phosphorylation of enzymes by protein kinases and the dephosphorylation by protein phosphatases. Other than in the case of the allosteric influence of adenine nucleotides on enzymes, a memory effect occurs here: after covalent phosphorylation of an enzyme the activity of that enzyme remains elevated or reduced, even if, subsequently, the ATP level drops. Thus, the modification of the enzyme activity is irreversible in this sense. On the other hand, the modification can be removed by the action of the phosphatase. It can be shown that for such a "controlled" control mechanism it is essential that in the modification-demodification cycle free energy is dissipated.

VII. THE DIFFERENT MEASURES OF THE CELLULAR ENERGY STATE

A. Free-Energy Differences

As can be gleaned from Figure 1, the free energy of hydrolysis of ATP (ΔG_p), the electrochemical potential difference for protons across the inner mitochondrial membrane and the electric potential difference across the plasma membrane are central free energies in cellular metabolism. One may thus propose to measure one of these and use the result as the criterion for the cellular energy "state". It is reasonable to use free-energy differences as a measure of the "energy state", because it is the magnitude of these that determine (1) the direction in which reactions run, or, in other words, whether they can run at all in the desired direction and (2) the rate of the reaction (though not exclusively, see above). Even from this point of view however, there are a number of problems with this approach. The first is subcompartmentation. Both inside the mitochondria and in the cytosol, there is a free energy of hydrolysis of ATP. Because the concentrations of ATP, ADP, and phosphate differ, these two ΔG_p's are generally different.[28] Also, their difference is not constant, or not even strictly related to another free energy difference, such as $\Delta \bar{\mu}_H$, because the adenine nucleotide translocator may be out of equilibrium to various extents depending on the metabolic state of the cell.[29] Even with respect to the proton electrochemical potential difference across the inner mitochondrial membrane, such a subcompartmentation may occur[30] with rather deleterious results for the possibility to use $\Delta \bar{\mu}_H$ as a measure of the free-energy state of the mitochondrion.[31] Second, we would have to decide which of the three (or more) free energies to consider as *the* measure of "the" energy state of the cell. Only if the reactions that convert the candidates (ΔG_p, $\Delta \bar{\mu}_H$, and the electric potential difference across the plasma membrane), i.e., the mitochondrial $\overrightarrow{H}^+$-ATPase plus the adenine nucleotide and phosphate translocations and the $\overrightarrow{Na}^+, \overleftarrow{K}^+$-ATPase would be in equilibrium, the three would be uniquely related. Since these systems are generally not all near equilibrium, one may observe an increase in ΔG_p simultaneously with a decrease in $\Delta \bar{\mu}_H$ (e.g., when both the adenine nucleotide translocator and the $\overrightarrow{H}^+$-ATPase are activated) under some conditions, whereas under other conditions ΔG_p and $\Delta\bar{\mu}_H$ may increase simultaneously (e.g., upon activation of the dehydrogenases).

I would tend to conclude that, essentially, there is no such thing as "the" energetic potential, or "the" energy state, of the cell. A cell has many different free-energy potentials. Indeed, this is an important tool for the cell in the regulation of its metabolism. For instance, if $\Delta\bar{\mu}_H$ and ΔG_p were at equilibrium, there would be little flux from the former to the latter and little ATP would be synthesized. This is yet another of those cases where we would wish to be able to describe biological systems in a simple fashion, but we cannot succeed because the biological system is complex in its very essence.

Of course all of this should not be taken to mean that the measurement of the intracellular phosphate potential is not important. On the contrary, the result, if properly corrected for compartmentation, will increase our insight into cellular energy metabolism. However, more is needed to obtain the complete picture.

B. Capacitances

Free-energy differences of reactions are potentials in the sense that they are defined as the change in free energy of the system when an infinitesimal amount of reaction is allowed to occur. Thermodynamics shows that these properties are important because they decide whether reactions will run in the desired direction and, to some extent, with what rate they will run. In actual practice however, the amounts of ATP used by proceeses are not infinitesimal. For instance, Akkerman and colleagues[2] showed that secretory stimulation of blood platelets by thrombin led to a 50% depletion of the cellular ATP content within 10 sec if the ATP regenerating processes had been inhibited. More specifically for secretion one finds the phenomenon that the number of ATP molecules hydrolyzed per single "molecular" event (which is the secretion of one granule per cell) is 36 million divided by the number of granules that are excreted per cell, which probably is much more than the one or two molecules of ATP that is consumed per usual metabolic event.

In such a context one may ask how much ATP the cell has available to support such ATP-consuming events. Rather than asking for the energy potential, this asks for the energy capacitance or energy storage in the cell. Since the amount of ATP per cell can be measured relatively simply, it would seem that this energy capacitance parameter may be a useful indication for the fitness of the cell to carry out certain processes. However, here also there are several *caveats*. The first is that the amount of ATP a cell can hydrolyze in a given activity is not simply given by the amount of ATP present in the cell at the time the activity is started. Since metabolism is a dynamic process, ATP is synthesized and degraded all the time. If there is a sudden increase in ATP consumption because a cellular activity is set into motion, then the amount of ATP available for all processes including the one just set into motion is equal to the amount of ATP present plus the amount of ATP that will continue to be synthesized subsequently. Subsequent to the drop in ATP concentration (or ΔG_p) many of the processes that were consuming the ATP that was being synthesized, may slow down, leaving more of the synthesized ATP for the new process. Also, and very importantly, the drop in ΔG_p may activate the processes that produce ATP. Indeed such "feedback" regulations are ample in cellular metabolism, the simplest and best-known being the stimulatory effect of a decrease in ΔG_p on mitochondrial oxidative phosphorylation.[32]

Consequently, the amount of ATP per cell is relevant only when processes are activated that consume ATP at a rate that is much higher than the maximum rate of cellular ATP production. In as far as I can estimate this from measurements of rates of ATP consumption[2,3] this situation is not yet obtained for instance when blood platelets secrete their storage granules.

Let us consider a secretory process that consumes ATP at a rate that does exceed the maximum rate at which the cell can produce ATP. In this case, the extent of secretion will be limited by the amount of ATP that the cell can mobilize rather than the amount of ATP the cell contains. This means that one should only count the measured ATP in as far as it can indeed be hydrolyzed by the processes that mediate secretion. For chromaffin cells, this means that one may not be allowed to include the large amount of ATP that is stored in the chromaffin granules. On the other hand, one should take into account the possibility that free energy may be stored in the cell in forms other than the hydrolytic free energy of ATP, which however, can readily be transformed into the latter. Atkinson[10] stressed this point and proposed to correct at least for the ATP equivalents that may be mobilized from ADP due to the adenylate kinase reaction:

$$2 \ ADP \rightleftharpoons ATP + AMP \qquad (20)$$

Because in most cells adenylate kinase is indeed very active, it is useful to use as parameter for "energy content", or "energy capacitance" of the cell the adenylate energy charge, rather than simply the ATP concentration:

$$\text{adenylate energy charge} = ([ATP] + \tfrac{1}{2}[ADP])/([ATP] + [ADP] + [AMP]) \qquad (21)$$

However, it should be realized that this only solves the problem of the free energy "stored" in the terminal anhydride bond of ADP and does not yet correct for other sorts of stored free energy. An example of a case where this correction may be insufficient is the significant variation in the concentration of fructose-1,6-diphosphate in blood platelets with the energetic work load imposed upon cellular metabolism.[33] Indeed, ATP "equivalents" present in this compound should be included in the calculation of the "total" energy charge. Other examples are the concentration of creatine phosphate in muscle cells, the free energy residing in Ca^{++} accumulated into the mitochondria in response to the electrical potential difference across that membrane, the free energy residing in the Na^+ (in bacteria this may be extremely relevant[34]) and K^+ gradients across the plasma membrane.

C. The Different Parameters for Energy "State" Tend to Correlate

After having read through all these reservations with respect to the utility of free-energy differences, or "energy charge" as parameters to characterize "the" energy state of the cell, how can we understand that sometimes, the use of such parameters seems to further our qualitative understanding of energy metabolism? Part of the answer may lie in correlations between all these parameters that are caused by two properties of cellular metabolism. First, in most cells the metabolic routes that synthesize or degrade adenine nucleotides tend to be orders of magnitude less active than the routes that interconvert them. As a consequence, it is possible to apply a "Frank-Condon principle" to cellular energy metabolism; in analyzing the energetics of the processes that interconvert ATP, ADP, and AMP one may assume that the sum of the concentrations of these compounds is constant. Only as a correction to this analysis, one may then analyze what happens to free-energy metabolism if this sum concentration slowly changes due to additional metabolism. Second, in most cells the adenylate kinase reaction is so active that the reaction may be very close to equilibrium. Together this gives us the following two relations between the three variables [ATP], [ADP], and [AMP]:

$$[ATP] + [ADP] + [AMP] = C_{adn} \qquad (22)$$

$$[ATP] \cdot [AMP] = K_{ak} \cdot [ADP]^2 \qquad (23)$$

Applying also the definition of the adenylate energy charge (Equation 21), and the definition

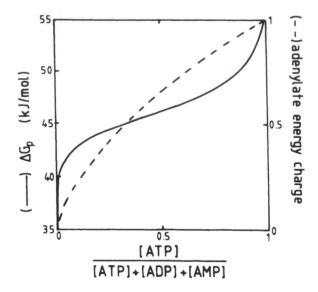

FIGURE 8. ATP concentration, phosphate potential (ΔG_p) and adenylate energy charge tend to be correlated. The adenylate kinase reaction was assumed to be at equilibrium (K_{eq} was taken equal to 1 for simplicity) and the sum of the adenine nucleotide concentrations was assumed to be constant. The standard free energy of hydrolysis of ATP was taken to be 28 kJ/mol and the phosphate concentration was taken to be constant at 1 mM.

of the phosphate potential (free energy of hydrolysis of ATP, $\Delta G_p = \Delta G_p^\phi + R \cdot T \cdot \ln\{[ATP]/([ADP] \cdot [phosphate]\})$ we can understand that these two properties are uniquely related to each other and to the concentration of ATP. That is, if a property correlates with [ATP], it will correlate with the adenylate energy charge and also with the phosphate potential.[35] (Of course, the three correlations will not have the same functional form, and will not necessarily be linear.) Figure 8 illustrates this.[10] Since the three parameters measure slightly different aspects of the energy state of the cell, it will be clear that the fact that they correlate has the consequence that any of the parameters will correlate with some sort of impressionist picture of the energetic competence of the cell.

VIII. CORRELATIONS BETWEEN ENERGY-STATE PARAMETERS AND CELLULAR ACTIVITIES

In Mosaic Non Equilibrium Thermodynamic analyses of a number of biological coupling systems, we found that one might quite often expect a linear correlation between the rate of the output process and the rate of the input process. Interestingly, the slope of this correlation would be an indication of the extent of coupling of the processes and of the mechanism of uncoupling that may be operative.[7] Experimentally such linear correlations have indeed been observed in quite a number of biological systems. One example is oxidative phosphorylation where respiratory rate and phosphorylation rate tend to vary linearly with each other as the work load is varied.[7,36] A second example is the variation of the rate of microbial growth with the rate of substrate consumption.[13] Interestingly, a similar such linear correlation is found in some secretory systems.[2,3]

When interpreting experimental data that reveal correlations between two properties, one often is caught in the dilemma that a correlation alone does not show which of the properties is the cause and which is the effect. As briefly alluded to above in discussing the characteristics of coupling, the analysis of two coupled processes may be even further complicated by the phenomenon that both tend to be cause and both tend to exhibit effects. Thus, if ATP hydrolysis is coupled to a free-energy-requiring process, then the rates of ATP hydrolysis will tend to correlate with the rate of that process for more than one reason: first, they will correlate because both will tend to increase when ΔG_p increases. I shall refer to this type of correlation as force-coupling correlation (see also Section IV.B). Second, they will correlate because, as the free-energy-requiring process proceeds, it will simply hydrolyze ATP. I shall call this correlation flux-coupling correlation. It may be noted that the flux-coupling correlation will be independent of the way in which the fluxes are varied, provided that the coupling between the two fluxes is not affected. On the other hand, two fluxes responding to changes in the same free-energy difference, may do that to a quite different extent (especially if the processes are not close to equilibrium).[7]

An interesting case of the flux-coupling correlation has been observed in thrombin-induced secretion by blood platelets.[2,3] The correlation between ATP consumption rate and secretion was similar, independent of whether temperature, the ATP content of the cell, or thrombin concentration was varied. Thus, this correlation probably reflects flux-coupling correlation. As a consequence, it is not necessary to invoke any force-sharing correlation to explain these results[3] although such a force-sharing correlation cannot of course be excluded.

IX. ENERGETICS AND FUSION

A. A Possible Role for Hydrostatic Pressure

An important step in exocytosis is the fusion of the intracellular storage vesicle with the plasma membrane of the secreting cell.[1] The thermodynamics of membrane fusion will be discussed in detail in a later chapter of this book. Here it may suffice to point out that among the thermodynamic driving forces in this type of fusion, there may be one that is rarely mentioned as a driving force of other metabolic processes. This is the hydrostatic pressure difference across the membrane of the secretory vesicles. Elsewhere[1] (and below) it has been reviewed that increasing the osmotic pressure of the extracellular medium inhibits this type of secretory fusion. Indeed, swelling of the secretory vesicles has been put forward as a prerequisite for fusion[14] and it is this part of the "chemiosmotic" hypotheses for exocytosis that seems to have survived recent tests.[15]

It may be instructive to see how osmotic swelling of the secretory vesicles may provide a driving force for fusion. In the absence of a temperature gradient across a membrane the transmembrane difference in chemical potential of the water is[7]

$$\Delta\mu_w = \overline{V}_w \cdot (\Delta P - \Delta\Pi) \tag{24}$$

Water permeability of biological membranes is extremely high, so that the chemical potential difference for water is zero at all times. This implies that if a compartment contains a high concentration of solute so as to generate a high osmotic pressure, water will flow into that compartment and swelling will occur. This swelling will have two consequences. First, the solute will be diluted, such that the osmotic pressure, Π, will dwindle. Second, as soon as the vesicle has swollen enough to attain the smallest surface-to-volume ratio possible (sphere), there is only one way left to increase the volume: the membrane will be stretched. This stretching will be opposed by an elastic force which will increase with the increase in volume of the vesicle, until the membrane ruptures, or until the pressure, P, exerted by this elastic force on the medium contained within the vesicle, becomes equal to the difference in osmotic pressure between inside and outside the vesicle.

If in some way, the vesicle can increase its volume, without having to further stretch its membrane, then this process will be driven by the driving force $\Delta\Pi \cdot dV$. dV corresponds

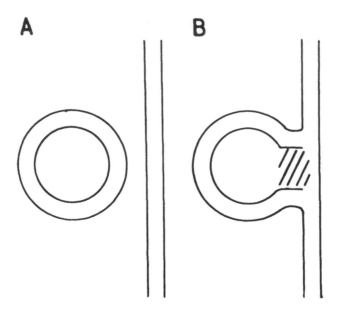

FIGURE 9.   The thermodynamics of the truly osmotic driving force for fusion. (A) gives the state before fusion and (B) gives a state just past the activated state complex. The shaded volume is gained by the inside of the vesicle upon the transition from state A to state B. If there is an osmotic pressure difference across the vesicle membrane, then this volume increase corresponds to a seizable thermodynamic driving force (see text).

to the increment in volume that occurs in the process. The transition between Figure 9A and B may be part of the fusion process. It is seen that in going from Figure 9A to Figure 9B, the vesicular content increases in volume by the shaded area. We may estimate the dimensions of the shaded volume as 10 nm by 2 nm by 2 nm. Assuming an osmotic pressure difference across the vesicular membrane corresponding to a concentration difference of 1 M, one can calculate that the transition between states A and B in Figure 9 dissipates a free energy of some $1 \cdot 10^{-19}$ J. For comparison, the free energy "content" of an ATP molecule amounts to some $8 \cdot 10^{-20}$ J: the amount of free-energy dissipated by this process is indeed of sufficient magnitude to be potentially relevant for the process.

B. A Possible Mechanism for a Modulating Role of the Transmembrane Electric Potential Difference in Fusion

Kamp and colleagues[37] have recently calculated the profile of the transmembrane electric potential resulting from the pumping of ions across a membrane. One of the results was that, especially if the surface of the membrane would not be extremely accessible to the major ions in solution, part of the transmembrane drop in electric potential may extend to the aqueous interphases bordering the membrane (see Figure 10A). For a Ca^{++} ion close to the membrane this would have the important consequence that it would be repelled by the membrane if it were present in the compartment where the potential is negative. Depolarization of the membrane would reduce this repulsion and might thus induce Ca^{++} binding to the lipids. In such a scenario, depolarization of the plasma membrane potential may lead to electrostatic adsorption of Ca^{++} to the inner side of the plasma membrane, where it might start fulfilling its role of perturbing the lipid bilayer structure, anticipating fusion with the apposed vesicle.

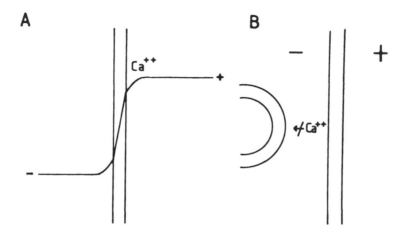

FIGURE 10. Modulation of fusion by the electric potential difference across the plasma membrane. (A) The profile of the electric potential across a membrane due to ion pumping.[37] Because part of the potential drop occurs in the interface, Ca^{++} may be stuck to the, paradoxically, positive side of the membrane. (B) A diagram illustrating how depolarization of the plasma membrane may stimulate fusion. Because in this case the positive potential is on the right (outside), i.e., *trans* to where Ca^{++} acts, Ca^{++} will be repelled by the membrane. If depolarization occurs, this repulsion will decrease and may allow Ca^{++} to bind to the inner side of the plasma membrane and start its fusogenic action.

X. CONCLUSION

Above, some general philosophies, kinetics and thermodynamics of cellular energy metabolism were discussed. To summarize these discussions one might say that free energies influence the direction and rates of a plethora of cellular processes. This influence can either be through serving as thermodynamic driving force, by catalyzing a "slow" step in a sequence of reactions, or by (in)activating a pathway. Cellular energy metabolism is too multifaceted to be summarized in terms of single parameters as ATP content, phosphate potential, or energy charge although we can understand why such summaries can be of help in a birds eye view of metabolism.

Obvious points of contact between secretory and energy metabolism, are the synthesis of the substances to be secreted, their transport into secretory vesicles, the transport of the vesicles (with the possible involvement of cytoskeletal components[38]) and the fusion of the vesicles with the plasma membrane. In the subsequent chapters of this book, the reader will find an exciting account of what we know, what we do not know but suspect, and what we do not know and have no idea about either, in this field.

ACKNOWLEDGMENT

I thank Gemma Kuijpers, Dean Astumian, and Frits Kamp for discussions. This study was supported in part by the Netherlands Organization for the Advancement of Pure Research.

REFERENCES

1. **Pollard, H. B., Ornberg, R., Levine, M., Kelner, K., Morita, K., Levine, R., Forsberg, E., Brocklehurst, K. W., Duong, L., Lelkes, P. I., Heldman, E., and Youdim, M.,** Hormone secretion by exocytosis with emphasis on information from the chromaffin cell system, *Vitam. Horm. (N.Y.)*, 42, 109, 1985.

2. **Akkerman, J. W. N., Gorter, G., Schrama, L., and Holmsen, H.,** A novel technique for rapid determination of energy consumption in platelets, *Biochem. J.*, 210, 145, 1983.

3. **Verhoeven, A. J. M., Mommersteeg, M. E., and Akkerman, J. W. N.,** Quantification of energy consumption in platelets during thombin-induced aggregation and secretion, *Biochem. J.*, 221, 777, 1984.

4. **Morowitz, H. J.,** *Foundations of Bioenergetics*, Academic Press, New York, 1978.

5. **Lehninger, A. L.,** *Biochemistry*, Worth, New York, 1975.

6. **Nicholls, D. G.,** *Bioenergetics*, Academic Press, New York, 1982.

7. **Westerhoff, H. V. and van Dam, K.,** *Thermodynamics and Control of Biological Free-Energy Transduction*, Elsevier, Amsterdam, 1987.

8. **Stouthamer, A. H.,** *Int. Rev. Biochem.*, 21, 1, 1979.

9. **Roels, J. A.,** *Energetics and Kinetics in Biotechnology*, Elsevier, Amsterdam, 1983.

10. **Atkinson, D. E.,** *Cellular Energy Metabolism and its Regulation*, Academic Press, New York, 1977.

11. **Van der Meer, R., Westerhoff, H. V., and Van Dam, K.,** Linear relation between rate and thermodynamic force in enzyme-catalyzed reactions, *Biochim. Biophys. Acta*, 591, 488, 1980.

12. **Bent, H. A.,** *The Second Law*, Oxford University Press, New York, 1965.

13. **Westerhoff, H. V., Lolkema, J. S., Otto, R., and Hellingwerf, K. J.,** Thermodynamics of growth. Non equilibrium thermodynamics of bacterial growth, the phenomenological and the mosaic approach, *Biochim. Biophys. Acta*, 683, 181, 1982.

14. **Pollard, H. B., Pazoles, C. J., Creutz, G. E., and Zinder, O.,** The chromaffin granule and possible mechanisms of exocytosis, *Int. Rev. Cytol.*, 58, 159, 1978.

15. **Baker, P. F. and Knight, D. E.,** Chemiosmotic hypotheses of exocytosis: a critique, *Biosci. Rep.*, 4, 285, 1984.

16. **Keizer, J.,** Thermodynamic coupling in chemical reactions, *J. Theoret. Biol.*, 49, 323, 1975.

17. **Kedem, O. and Caplan, S. R.,** Degree of coupling and its relation to efficiency of energy conversion, *Trans. Faraday Soc.*, 21, 1897, 1965.

18. **Onsager, L.,** Reciprocal relations in irreversible processes. I, *Physiol. Rev.*, 37, 1, 1931.

19. **King, E. L. and Altman, C.,** A schematic method of deriving the rate laws for enzyme-catalyzed reactions, *J. Physiol. Chem.*, 60, 1375, 1956.

20. **Hill, T. L.,** *Free Energy Transduction in Biology*, Academic Press, New York, 1977.

21. **Malmström, B. G.,** Cytochrome c oxidase as a proton pump. A transition state mechanism, *Biochim. Biophys. Acta*, 811, 1, 1985.

22. **Boyer, P. D.,** Correlations of the binding change mechanism with a new concept for proton translocation and energy transmission, in H^+-*ATPase (ATP Synthase): Structure, Function, Biogenesis*, Papa, S., Altendorf, K., Ernster, L., and Packer, L., Eds., Adriatica Editrice, Bari, Italy, 1985, 329.

23. **Westerhoff, H. V. and Astumian, R. D.,** The dynamics of electrostatic interactions between membrane proteins, in *Towards a Cellular Enzymology*, Klyosov, A., Varfolomeev, S., and Welch, G. R., Eds., Plenum Press, New York, in press.

24. **Glasstone, S., Laidler, K. J., and Eyring, H.,** *The Theory of Rate Processes*, McGraw-Hill, New York, 1941.

25. **Pietrobon, D. and Caplan, S. R.,** Flow-force relationships for a six-state proton pump model: intrinsic uncoupling, kinetic inequivalence of input and output forces, and domain of approximate linearity, *Biochemistry*, 24, 5764, 1985.

26. **Westerhoff, H. V., Tsong, T. Y., Chock, P. B., Chen, Y., and Astumian, R. D.,** How enzymes can capture and transmit free energy from an oscillating electric field, *Proc. Natl. Acad. Sci. U.S.A.*, 83, 4734, 1986.

27. **Astumian, R. D., Chock, P. B., Tsong, T. Y., Chen, Y., and Westerhoff, H. V.,** Can free energy be transduced from electrical noise?, *Proc. Natl. Acad. Sci. U.S.A.*, 84, 434.

28. **Lanoue, K. F. and Schoolwerth, A. C.,** Metabolite transport in mammalian mitochondria, in *Bioenergetics*, Ernster, L., Ed., Elsevier, Amsterdam, 1984.

29. **Wanders, R. J. A., Groen, A. K., Meijer, A. J., and Tager, J. M.,** Determination of the free-energy difference of the adenine nucleotide translocator reaction in rat-liver mitochondria using intra- and extra-mitochondrial ATP-utilizing reactions, *FEBS Lett.*, 132, 201, 1981.

30. **Westerhoff, H. V., Melandri, B. A., Venturoli, G., Azzone, G. F., and Kell, D. B.,** A minimal hypothesis for membrane-linked free-energy transduction. The role of independent, small coupling units, *Biochim. Biophys. Acta*, 768, 257, 1984.

31. **Westerhoff, H. V. and Chen, Y.,** Stochastic free-energy transduction, *Proc. Natl. Acad. Sci. U.S.A.,* 82, 3222, 1985.
32. **Chance, B. and Williams, G. R.,** Respiratory enzymes in oxidative phosphorylation. III. The steady state, *J. Biol. Chem.,* 217, 409, 1955.
33. **Akkerman, J. W. N., Driver, H. A., Dangelmaier, C. A., and Holmsen, H.,** Alterations in ^{32}P-labelled intermediates during flux activation of human platelet glycolysis, *Biochim. Biophys. Acta,* 805, 221, 1984.
34. **Skulachev, V. P.,** Sodium bioenergetics, *Trends Biochem. Sci.,* 9, 483, 1984.
35. **Rottenberg, H.,** An irreversible thermodynamic approach to energy coupling in mitochondria and chloroplasts, in *Progress in Surface and Membrane Science,* Vol. 12, Cadenhead, D. H. and Danielli, J. F., Eds., Academic Press, New York, 1978, 245.
36. **Van Dam, K., Westerhoff, H. V., Krab, K., Van der Meer, R., and Arents, J. C.,** Relationship between chemiosmotic flows and thermodynamic forces in oxidative phosphorylation, *Biochim. Biophys. Acta,* 591, 240, 1980.
37. **Kamp, F., Chen, Y., and Westerhoff, H. V.,** Spatial charge distribution and electric potential profile across energy coupling membranes, *Biophys. Chem.,* submitted.
38. **Vale, R. D., Schnapp, B. J., Reese, T. S., and Sheetz, M. P.,** Organelle, bead, and microtubule translocations promoted by soluble factors from the squid giant axon, *Cell,* 40, 559, 1985.

Prestimulation Events

Chapter 3

BIOSYNTHESIS AND STORAGE OF SECRETORY PROTEINS

B. Rossignol, G. Herman, and P. Mauduit

TABLE OF CONTENTS

I. INTRODUCTION

Over the last 20 years, our knowledge of the mechanisms involved in the biosynthesis, intracellular transport, storage, and release of secreted proteins has progressed considerably. The period 1950 to 1965 saw the description of cellular architecture, the first studies of the radioactive labeling of proteins, and the development of subcellular fractionation. Building on these advances, Palade, by virtue of a combination of microscopy and biochemical techniques, was able to establish a general scheme for protein secretion. In his Nobel lecture in 1974, Palade outlined the six principal stages of the secretory process as follows.[1]

In the first stage, proteins destined for export are synthesized on polyribosomes bound to the endoplasmic reticulum. A signal peptide is believed to play a role in the attachment of the polysomes to the reticulum membrane.[2] During the second stage, the newly synthesized proteins undergo vectorial transport from the large subunit of ribosome, which remains attached to the membrane of the reticulum, to the lumen. This segregation[3] appears to be irreversible. The protein now assumes a globular configuration and is subject to post-translational modifications under the influence of several enzyme systems (formation of disulfide bridges,[4] preliminary glycosylation,[5] hydroxylation[6]). The third stage comprises the intracellular transport of the proteins which are transferred from the endoplasmic reticulum to the *cis*-side of the Golgi apparatus by transitional vesicles.[7] This step requires energy to be supplied in the form of ATP.[8] It is the first "energy dependent lock" that controls the transport of secretory products. The fourth stage refers to the concentration of secretory products in the condensing vacuoles on the *trans*-side of the Golgi.[9] This process is not energy dependent. The intervention of sulfated polyanion at this stage has been proposed.[10] Certain post-translational modifications also occur in the Golgi region (terminal glycosylation,[11] sulfation[12] and some proteolysis may begin at this level). The fifth stage consists of the packaging of the protein in secretory granules. Different proteins may be found together in the same granule, or in specific granules, depending on the type of tissue. The sixth and final stage is the release of the proteins by fusion between the membrane of the granule and a defined region of the plasma membrane. This energy-dependent step represents the "second energy dependent lock".[1]

This general outline seems to apply, in broad terms, to all types of protein secretion; however, some differences emerge at certain stages. Hence, in endocrine glands, exocytosis can take place over the whole surface of the plasma membrane. The storage phase is bypassed in cells which secrete their products constitutively and are not subject to discharge regulation by an external stimulus. Intracellular transport may be "diverted" for example, in macrophages, secretory products seem to be transferred from the Golgi to endocytotic vacuoles.

Since these basic principles were laid down, our knowledge of biochemical events at the ultrastructural level has advanced, and consequently new data begin to provide some answers to different questions.

- Can the two classes of secretion, constitutive and stimulus dependent, co-exist in the same cell type? If so, is the fabrication of proteins similar in the two pathways?
- How are proteins destined for export and those destined for membranes or lysosomes sorted one from another? Do post-translational modifications act as an "address" for the protein?
- Is the "processing" of a particular protein specific for a given cell type, or can this take place in any cell?
- Is the cytoskeleton involved in intracellular transport? Are intracellular transport and "processing" of proteins also subject to (external) regulation?

The secreted proteins of multicellular organisms present an extremely wide variety of structures and biological functions. They range from antibodies to proteoglycans of the extracellular matrix passing through plasma proteins, digestive tract enzymes, and hormones. As a result, this is an extensive field, and in the first part of this chapter, we will attempt to summarize the biochemical mechanisms for the events occurring in each of the different cellular compartments, namely, the endoplasmic reticulum, Golgi apparatus, condensing vacuoles, and secretory granules. In the second part, we will show how contemporary data can begin to answer the questions posed above.

II. BIOSYNTHESIS AND POST-TRANSLATIONAL MODIFICATIONS OF SECRETED PROTEINS

A. "Rough Endoplasmic Reticulum Stage

1. Biosynthesis of Polypeptide Chains and their Translocation into the Cavity of the Reticulum

The mechanisms of protein fabrication at the level of the rough endoplasmic reticulum (RER) have been the subject of numerous investigations, which have been frequently reviewed.[1,13-17] The broad outlines are common to proteins destined for export,[1] for membranes,[15,18] or for lysosomes.[19] In this section, we give an account of the general mechanisms of polypeptide chain synthesis on bound ribosomes and the movement of these chains across the reticulum membrane. The development of in vitro experimental systems and genetic engineering technology have permitted an overall scheme for this process to be established. It has long been known that ribosomes are capable of binding to the reticulum membrane and that bound ribosomes are able to exchange subunits with free ribosomes with each cycle of protein synthesis.[15,20] A further observation was that ribosomes bound in this way, but devoid of nascent polypeptide chains could be liberated by incubation in medium containing a high salt concentration; in contrast, treatment with puromycin (which detached the growing chain) was necessary to release ribosomes bearing a nascent chain.[21,22] These early experiments show that the emerging chain plays a role in the binding of the polysome to the membrane. Furthermore, once the polypeptide chain is formed at the level of the reticulum, it rapidly becomes resistant to proteases.[23] Milstein et al.[2] formulated the signal hypothesis, based on the observation that the light chains of immunoglobulins were synthesized in the form of a precursor which lost a segment from the amino terminal to give the mature light chain. Other data mostly obtained from cell-free systems, supported the hypothesis of a signal peptide.[23-27] To summarize these results: the products formed in the absence of reticulum membranes contained a terminal segment (called the signal sequence) which does not form part of the final secreted product found in the lumen of the reticulum. On the other hand, cytoplasmic proteins do not contain this sequence, which must therefore contain the information necessary to allow binding of the polysome to the membrane and translocation of the polypeptide chain to the intravesicular space. This translocation was quickly perceived to be concomitant with translation. Nowadays, the term "cotranslational translocation" is used for the case of secreted and lysosomal proteins, and "cotranslational integration" for proteins destined for insertion into membranes.[17]

The whole process may be divided into functional stages: the first comprising the recognition and binding of polysomes to the reticulum membrane and the second the translocation of the polypeptide chain.

Two main components involved in the first stage have been purified from the microsomal fractions.[16,17,28-36] The first of these is the "signal recognition particle" (SRP), an 11 S cytoplasmic ribonucleoprotein. The particle is composed of six polypeptide chains of widely differing molecular weights (72, 68, 54, 19, 13, and 9 kdaltons) and a molecule of RNA of about 300 nucleotides. All these components are essential to the activity of the SRP. SRP has the properties of an extrinsic membrane protein and, at physiological salt concentrations, exists in an equilibrium between a form bound to the reticulum membrane, a free form, and a polysome-associated form.[35] The second component is an SRP receptor, found on the reticulum membrane, which is also called the "docking protein". The SRP receptor is an intrinsic membrane protein of 72 kdaltons and was purified from detergent extracts of membranes by affinity chromatography. It possesses two domains, one within the membrane and the other on the cytoplasmic side.[36] In addition to these two components, a number of membrane constituents have roles in the binding of polysomes and membranes, and among these, ribophorins I and II, which at least stabilize the binding to the membrane, should be mentioned.[37]

The functions of these different components can be summarized as follows. When a messenger RNA coding for a protein which includes the signal sequence is being translated, the affinity of SRP for the translating ribosome increases by several orders of magnitude. The result is that SRP binds to the ribosome-signal sequence complex by virtue of its recognition of the signal peptide. Hence, SRP blocks translation immediately after the formation of the signal sequence.[16] This occurs while the ribosome is still free in the cytoplasm. The complex SRP-ribosome-peptide then binds to the reticulum membrane and the inhibition of translation is revoked by interaction with the docking protein. Chain elongation resumes and the protein can now be transferred across the membrane. Once the SRP ribosome complex has been recognized by the SRP receptor, other proteins, including ribophorins I and II, seem to intervene in the interaction between the ribosome and the membrane. However, to date, the observations have only been made in vitro, and it is difficult to speculate on the mode of action of SRP in vivo.

The second functional stage includes the translocation and the subsequent modification which the protein may undergo.

It has been stated that translocation comes about through a "pore-like" structure. This is simply meant to indicate that the permeability barrier presented by the membrane is transiently lifted to allow the nascent chain to pass through. Several mechanisms have been proposed to explain the penetration of the membrane by the signal sequence.[14,15,38,39] Among these, one is that the polypeptide forms a loop. The N-terminal residue remains on the cytoplasmic side, the loop crosses the membrane, and the nascent protein appears in the lumen of the reticulum. Before the polypeptide chain is complete, the signal portion may be removed by an endopeptidase situated on the luminal side of the cavity. The fate of the peptide signal is unknown, however we can suppose that it is rapidly degraded. The peptidase has no specificity for the amino acid on the N-terminal side of the peptide bond, however, it requires a short-side chain amino acid (glycine, serine, alanine, or cysteine) on the C-other terminal side.[14,38] The name "transport peptide" has been suggested for this sequence, since this corresponds more closely to its function.[38,39] Ovalbumin is a case in point, where the "transport peptide" is situated in the middle of the polypeptide chain and is not excised.[38,40,41] The structure of the transport peptide is variable[38,39] but always has a large proportion of hydrophobic residues. The passage of the protein into the cavity of the reticulum is endergonic, and several theories as to the origin of the energy input have been put forward, for example an interaction between the peptide and phospholipids or a conformational change in a protein at the point of contact between the ribosome and the membrane.[42-44] This debate was the subject of a recent review.[45]

The protein undergoes a series of transformations during its translocation and its stay in the lumen of the reticulum. These will be described in the following section.

2. Post-Translational Modifications

We will treat three types of post-translational modification taking place in the rough endoplasmic reticulum: glycosylation, hydroxylation of proline and lysine residues, and the formation of disulfide bridges.

a. Glycosylation

The mechanism of protein glycosylation has already been extensively reviewed.[46-49] and has inspired a great deal of research at the level of the biochemical pathways involved as well as the ultrastructural level. Without wishing to resort to hasty generalizations, in this paragraph, we select and describe the basic common pathways, and provide the reader with information which will allow him to pursue the subject further. Glycoproteins contain both N- and O-glycosidic linkages between the polypeptide chain and the carbohydrate moiety. It is now recognized that N-glycosidic links between asparagine and oligosaccharides are formed at the level of the RER. These oligosaccharides vary considerably in structure,[47] but possess a common core sequence at the protein-oligosaccharide linkage shown in Figure 1. This core is modified to different extents by the addition of further sugar residues. Thus, the polysaccharide moiety may consist entirely of mannose or complex structures containing mannose, N-acetylglucosamine, galactose, sialic acid, and fucose. The assembly of the polysaccharide begins with a common oligosaccharide precursor (Figure 2) which is built up on a lipid (dolichol) in the reticulum. This oligosaccharide is transferred to the protein and then modified; however, galactose, sialic acid, and fucose are not added before the protein reaches the Golgi region. The donors for oligosaccharide synthesis are either (in the case of the sugars indicated by the box in Figure 2) nucleotide sugars, or, for the rest, dolichol-phosphate-sugars. Although it has been clearly established that the transfer of the completed oligosaccharide to an aspargine of the protein occurs in the lumen of the reticulum, the assembly of this oligosaccharide on the dolichol from its various precursors is less well understood. Experiments carried out using lectins have made it possible to locate some intermediates in the formation of this precursor oligosaccharide.[50,51] It has been shown that an intermediate consisting of mannose and N-acetylglucosamine is found on the cytoplasmic side of the RER, while the completed oligosaccharide is situated in the interior of the reticulum. This suggests that a part of the oligosaccharide is synthesized on the cytoplasmic side and that the necessary components, the nucleotide sugars GDP-mannose and UDP N-acetylglucosamine do not cross the RER membrane. The growing oligosaccharide would then traverse the membrane to the internal face, where it would be completed from dolichol phosphate presursors, dolichol-P-glucose and dolichol-P-N-acetylglucosamine; and finally transferred from the dolichol carrier to the protein.[52] It has also been suggested that these dolichol phosphate sugar precursors are formed, by the reaction:

$$XDP\text{-sugar} + Dol\text{-}P \rightarrow Dol\text{-}P\text{-sugar} + XDP$$

on the cytoplasmic side and subsequently transferred to the internal face of the reticulum.[53] What factors control the transfer of the oligosaccharide to the protein? In vitro, the presence of three glucose residues facilitates this process, but does not appear to be indispensable. Various experiments carried out in the absence of glucose[54-56] nevertheless indicate an important role for this sugar. An interesting result was obtained with thyroid slices, where the depletion of energy reserves led to the accumulation of an oligosaccharide intermediate lacking glucose and to a reduction in protein glycosylation.[57] The sequence of the peptide acceptor also appears to be important.[58] In general, this sequence is Asn-X-Ser/Thr, where X can be any amino acid except proline or aspartic acid. The elongation of the peptide chain favors the addition of carbohydrates as does a β-sheet configuration in the "transfer zone". The formation of a hydrogen bond between the amino group of asparagine and the hydroxyl group of serine or threonine has been proposed.[58] This would render the amide more nucleophilic and therefore more reactive with the sugar donor. These data demonstrate that the glycosyltransferase which catalyzes the oligosaccharide transfer must recognize a specific protein structure. According to the hypothesis that carbohydrate transfer occurs during the passage of the protein into the lumen of the reticulum, these conditions will not always be

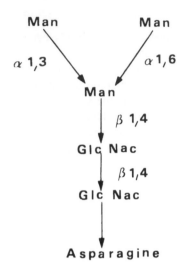

FIGURE 1.   Glycosylation.

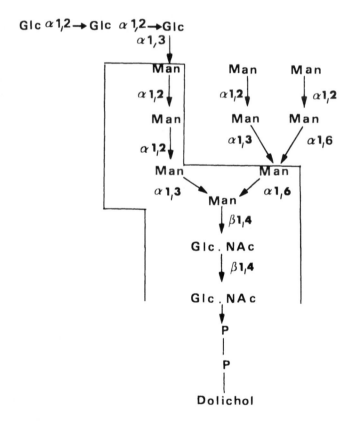

FIGURE 2.   Polysaccharide assembly on dolichol.

simultaneously fulfilled, hence not every Asn-X-Ser/Thr sequence will be glycosylated. All the evidence points to the conclusion that the glycosyltransferase requires particular structures for both the dolichol-bound oligosaccharide and the acceptor sequence on the protein to be present at the same time.[60]

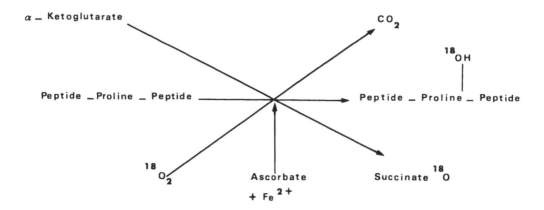

FIGURE 3.   Hydroxylation of proline residues.

i. Processing of the Oligosaccharide in the Reticulum

Once added to the protein, the oligosaccharide undergoes a number of modifications.

The carbohydrate structure shown in Figure 1 is transferred to an asparagine residue of the nascent protein during its vectorial transport towards the lumen of the reticulum. Processing begins with the excision of the terminal glucose residue by a 1,2-glucosidase, then the two remaining glucoses are removed by a 1,3-glucosidase. Finally, at least one mannose residue is removed by a mannosidase. Depending on the glycoprotein, this processing may take place either during the translocation of the protein after the attachment of the oligosaccharide, or after the passage of the protein into the lumen of the reticulum.[61,62]

b. Hydroxylation of Proline and Lysine Residues

The constituents of the extracellular matrix, including collagen, are formed and secreted by a process similar to that described above. Information concerning the structure and biosynthesis of these components may be found in several reviews.[63-65] The polypeptide chains of collagen are built up on bound ribosomes and are known as "pro-α" chains. These precursor molecules possess "extension peptides at the N and C terminals" and also a large number of proline and lysine residues which are hydroxylated in the cavity of the reticulum. The hydroxylation of proline seems to be important since inhibition of this reaction prevents the formation of the procollagen helix from several pro-α chains, a process which also takes place in the RER. In addition, lysine hydroxylation is essential for subsequent glycosylation, galactose, and then glucose residues being attached to the hydroxyl lysine.[63,67]

Hence, hydroxylation must occur before the association of pro–α chains and glycosylation. The proline (and lysine) hydroxylase has been revealed by the binding of peroxidase-conjugated antibodies to reside in the rough endoplasmic reticulum.[68] This enzyme requires Fe^{2+} and molecular oxygen. After incubating a tissue in the absence of oxygen in an iron-chelating medium, one observes an accumulation of protocollagen, the nonhydroxylated form of collagen. This inhibition is reversible. The hydroxylation reaction was described some time ago.[69] and its overall mechanism is summarized in Figure 3.

Hydroxylysine also plays a role in the ultimate structure of the collagen in that it may be converted by an oxidase into a form capable of forming cross links between polypeptide[70] chains. Thus, the post-translational modifications of the primary structure of the collagen chains which occur in the reticulum seem to be a prerequisite for the later transformations which give rise to the collagen as it is found in the extracellular matrix.

c. Formation of Disulfide Bridges

The formation of disulfide bridges has already been reviewed.[71] In a model of translation in vitro, it has been shown that proteins synthesized in the rough endoplasmic reticulum rapidly acquire their conformation (protease-resistant) with the formation of disulfide bridges if oxidized glutathione (G-S-S-G) is added to the medium.[72] Hence, it is necessary to maintain a sufficient redox potential in the RER to allow the formation of disulfide bridges. On the other hand, a protein disulfide isomerase, capable of catalyzing this linkage has been demonstrated in the RER, and it seems likely that this enzyme plays a role in the reaction, rather than this being the result of a simple spontaneous oxidation. This enzyme is present in tissues secreting proteins which contain disulfide bridges and its level of expression correlates with the amount of synthesis of such proteins, for example, when stimulated B lymphocytes, which are producing immunoglobulins, are compared with nonstimulated B cells.[73]

There is also a strict correlation between the activities of proline hydroxylase and protein disulfide isomerase.[74] The properties of this latter enzyme (salt extraction, sensitivity to proteases) suggest that it is firmly associated with the RER membrane and may reside on the luminal side. It catalyzes the exchange of thiol disulfides on proteins. In the presence of a weak oxidizing agent (a mixture of GSSG and GSH) it favors the formation of native disulfide bridges within proteins. These results have been observed for serum albumin, immunoglobulins, and procollagen.[71] The enzyme has been purified from bovine[75] and rat[76,77] liver. It is a relatively abundant protein, representing up to 2% of the microsomal protein in rat liver.[71] At the moment, it is not known what position this enzyme activity occupies (either temporally or spatially) with respect to the other reactions taking place in the reticulum, such as glycosylation and cleavage of the signal peptide. In addition, the source of oxidizing equivalents is unclear.

d. Inhibitors of Post-Translational Modifications
i. Inhibition of Translocation and Hydrolysis of the "Transport Peptide"

The incorporation of a leucine analog, β-hydroxyleucine, renders the protein incapable of crossing the reticulum membrane.[78] This observation suggests the importance of the series of leucine residues in the transport peptide.[39,78,79]

Replacing the C-terminal residue of the transport peptide by an analog of threonine, for example, β-hydroxynorvaline, in the propeptide of prolactine causes inhibition of release of the signal peptide.[80] However, translocation still occurs, showing that the liberation of the leader peptide is not an obligatory step in the transport of secreted proteins into the reticulum cavity.

ii. Inhibition of Glycosylation, Processing of the Oligosaccharide, and Hydroxylation of Proline and Lysine

The inhibition of glycosylation and of oligosaccharide processing has been the subject of recent reviews.[81-83] A number of compounds may be used among these tunicamycin, streptovirudin, and the antibiotic 24010, which inhibit the formation of dolichyl-pyrophosphoryl *N*-acetylglucosamine. In some cases, proteins whose glycosylation is blocked in this fashion, are secreted and in others they are not.[82,83] Some analogs of D-glucose and D-mannose are also inhibitory, and possess antiviral activity against enveloped viruses.

Modifications of the polypeptide chain can also lead to an inhibition of glycosylation. Examples are the substitution of β-hydroxynorvaline for threonine in the Asn-X-Thr sequence and of β-fluoroasparagine for asparagine.[81]

Recently, several compounds which inhibit the processing of the oligosaccharide have been brought to light.[83] Among these, is deoxynojirimycin which inhibits glucosidases I and II and can therefore be used to study the importance of glucose release in the exit of nascent proteins from the RER.

As stated above, the hydroxylation of proline and lysine residues may be inhibited in vitro by iron chelating agents and the absence of molecular oxygen. This inhibition thus blocks the translocation of procollagen chains.

B. Golgi and Secretory Granule Stage

Proteins which have been synthesized and modified in the rough endoplasmic reticulum are transferred to the Golgi apparatus by means of small vesicles, which seem to act as a "shuttle" between these two organelles. This transport has been described in detail in several reviews.[1,48,84-86] This unidirectional transfer seems to be energy-dependent since inhibitors of oxidative phosphorylation and uncouplers cause a retention of secretory products in the reticulum.[8] It seems that, to quote Farquhar and Palade[86] "passage of secretory products through the Golgi complex is obligatory and involves extensive modification and transfer to a membrane container which is competent to fuse with the plasmalemma at the time of exocytosis. It is in this Golgi-derived membrane container that concentration of secretory products is accomplished, but concentration is not an obligatory operation." At the level of the Golgi apparatus, the proteins undergo further modifications, principally glycosylation, sulfation, phosphorylation, and the first steps of proteolysis.

1. Structure of the Golgi Apparatus

A complete account of the structure of the Golgi complex may be found in a number of review articles.[48,85,86] Briefly, this organelle consists of stacked cisternae of varying compositions. The stack has polarity, with the *cis*-side adjacent to the reticulum and the other, *trans*-side, oriented towards the secretory granules or the centriole. Several different areas with the Golgi may be distinguished by the distribution of certain enzyme activities. Thiamine pyrophosphatase (TPPase),[87] acid phosphatase,[87] and galactosyl-transferase[88] are found in the *trans* cisternae. Nicotinamide dinucleotide phosphatase (NADPase)[89] and *N*-acetylglucosamyl transferase I[90] are located in the intermediate cisternae of the stack, whereas the *cis* region is rich in strong reducing agents, as evidenced by the reduction of osmium tetroxide.[91]

Several models have been put forward to explain the transfer of material through the Golgi region. Among these are the "membrane flow" or "cisternal progression" model,[92] the "distillation tower" model,[93] and the "stationary cisternae" model.[85,86] This latter scheme, which is now widely accepted, supposes that only the secretory products move through the stack while the cisternae remain fixed.

Transfer between the different elements of the stack is assumed to occur by means of vesicles, mainly at the dilated extremities of the Golgi cisternae.[85,86]

Monensin, a carboxylic acid ionophore capable of exchanging Na^+ or K^+ ions for protons, has been used extensively for studying transport within the Golgi and the role of the various subcompartments.[48,85,94] This agent destroys Na^+ and K^+ gradients within the cell, accompanied by a dilatation of the Golgi cisternae and a slowing down of intracellular transport. A number of biochemical processes (for example, protein synthesis) are unaffected by the ionophore, whose effects are reversible. It has been suggested[94] that monensin might interrupt intracellular traffic at the level of the intermediate cisternae of the Golgi. These disturbances might be due either to an alteration in the Na^+/K^+ balance in the cytoplasm[94] or to the existence of a proton gradient, maintained by a proton pump, across the membranes of the Golgi cisternae, the dissipation of which by the action of the ionophore, could lead to an increase in intracisternal pH.[85] Disruption of the microtubule network with colchicine has also been suggested to affect flow through the Golgi region (see below).

2. Modification of Secretory Products in the Golgi Apparatus and Secretory Granules
a. Glycosylation

Three types of glycosylation may be distinguished: the completion of N-glycosidic-linked oligosaccharides, the biosynthesis of oligosaccharides which are attached to the protein by O-glycosidic linkages, and the addition of glycosaminoglycans to proteoglycans (see Section III.B.2.b).

i. Completion of N-Linked Oligosaccharides

Several excellent reviews cover the series of reactions leading to the completion of these oligosaccharides;[47-49,85] therefore, we will merely summarize these events. The carbohydrate moiety of the glycoproteins arriving at the *cis*-side of the Golgi stack consists of nine mannose and two *N*-acetylglucosamine residues. The first step is the removal of between one and four mannose residues by mannosidase I, followed by the addition of an *N*-acetylglucosamine residue by *N*-acetylglucosaminyl transferase I. Two further mannose residues are cleaved by mannosidase II and two *N*-acetylglucosaminyl residues are joined to the oligosaccharide by *N*-acetylglucosaminyl transferases II and IV. Galactose, sialic acid, and fucose residues are subsequently added by specific transferases.

The different compartments of the Golgi apparatus have been separated by sucrose density gradient centrifugation and the processing enzymes localized in the various membrane fractions. In this way, mannosidase I has been found to be located in the *cis* region; *N*-acetylglucosaminyl transferases I and II, mannosidase II, and fucosyl transferase in the intermediate cisternae, and galactosyl- and sialyl transferases in the *trans* cisternae. Several of these enzymes have also been located *in situ* by cytochemical methods. Good correlations with the results of subcellular fractionation were obtained for galactosyl transferase,[88] sialyl transferase,[95] and *N*-acetylglucosaminyl transferase.[96] However, in hepatocytes, mannosidase II was found to be distributed throughout all the cisternae.[97] Thus, the enzymes involved in the processing of the oligosaccharide are arranged in the stack in a fashion which allows the sequential modification of the glycoprotein as it passes from the *cis* to the *trans* cisternae.

In the Golgi region, the sugar donors seem to be nucleotide sugars. During the transfer of galactose from UDP-galactose to the glycoprotein, UDP is liberated and may be degraded to UMP by a nucleotide diphosphatase (UDPase). The assay for this enzyme is identical to that for thiamine pyrophosphatase (TPAse) which might explain why this activity appears to be co-located with galactosyl transferase. The transfer of nucleotide sugars from the cytoplasm to the interior of the Golgi cisternae probably takes place by means of transporter-exchanging nucleotide sugars and nucleotide monophosphates.[98] Glycoproteins which have passed through the Golgi apparatus are resistant to endo-β-*N*-acetylglucosaminidase H (EndoH), which cleaves the oligosaccharide of proteins emerging from the endoplasmic reticulum. This property has often been exploited to demonstrate the transfer of proteins through the Golgi region.

ii. O-Glycosylation

The mechanisms involved in the formation of O-glycosidic linkages between *N*-acetylgalactosamine and a serine or threonine residue are less well understood than those for N-glycosidic links. The sugar is transferred directly from UDP-*N*-acetylgalactosamine without an intermediate lipid donor. Several pieces of experimental evidence point to the Golgi apparatus *cis* the site of this transfer, in particular the timing of this step with respect to others, and the colocalization of the enzyme(s) with other Golgi enzymes.[48,85] The assembly of the O-linked oligosaccharide seems to take place in several of the Golgi compartments.[99]

b. Proteoglycan Biosynthesis and Sulfation

Proteoglycans contain glycosaminoglycan chains, consisting of hexosamine and uronic acid residues linked to the protein by O-glycosidic links between a xylose and a serine or threonine residue. Although it is uncertain whether the transfer of xylose from UDP-xylose to the protein occurs in the rough endoplasmic reticulum or in the Golgi, it is believed that the elongation of the glycosaminoglycan chains takes place in the Golgi.[48,85] However, none of the enzymes involved has yet been localized *in situ*. It is also known that sulfation of the proteoglycans occurs in the Golgi.[12,48,85,86] The sulfate donor is adenosine-3'-phosphate-5'-phosphosulfate which might be transported to the interior of the Golgi cisternae by a system of an antiport.[98] It has been suggested that sulfation could be the last post-translational modification undergone by proteoglycans before their liberation outside the cell.[85,100]

It is interesting that the sulfation reaction is particularly sensitive to monensin. At a low monensin concentration ($10^{-8} M$) an inhibition of sulfation was observed while the elongation of the glycosaminoglycan chains was able to continue.[101] Also, at low concentrations the ionophore was able to distinguish between the sulfation of proteoglycans and the completion of N-linked oligosaccharides.[102]

c. Phosphorylation

Many studies have focused on the biosynthesis of lysosomal enzymes and the subject has been extensively reviewed.[48,85,86] The phosphorylation of mannose residues of lysosomal hydrolases seems to be a key reaction and proceeds in two stages.[85,103,104] First, N-acetyl-glucosamine-1-phosphate transferred from UDP-N-acetylglucosamine to a mannose residue on the glycoprotein with the formation of a phosphodiester bond. In the second step N-acetylglucosamine is cleaved by a phosphodiester glycosidase leaving the oligosaccharide-mannose-6-phosphate. These reactions take place in the *cis* region of the Golgi.[104] The mannose-6-phosphate acts as the signal which directs these glycoproteins towards the lysosomal compartment (see below).

d. Proteolysis

Several proteins including albumin, the membrane proteins of enveloped viruses, polypeptide hormones, and growth factors are synthesized in the form of precursors. These products undergo at least two proteolytic steps; the first removing the "pre" or signal peptide and the second leading to the formation of the mature protein with full biological activity. In some cases the prepropeptide is a common precursor for several proteins.

These post-translational modifications have been extensively reviewed.[48,86,105-108,113] In this section we will give an outline of the process illustrated by several specific examples. We have already covered the cleavage of the signal (or pre) sequence in the section dealing with events in the rough endoplasmic reticulum. The proteolysis of the propeptides may begin in the Golgi apparatus and continue in the secretory granules.

The cleavage site usually contains a pair of basic amino acids and the secondary and tertiary structures of the protein also appear to contribute to the specificity of proteolysis. The molecular weights of the propeptides can be extremely variable (for example 9 kdaltons for proinsulin compared to 55 kdaltons for proalbumin) as can the sizes of the discarded peptides. The use of cell-free systems translating mRNA in vitro and the sequencing of cDNA have allowed considerable advancement in our knowledge of the structure of protein precursors, particularly of hormones. One important discovery is that of propeptide sequences common to several proteins.

Insulin is synthesized as a prepropeptide and loses its signal peptide (of 23 amino acids) in the rough endoplasmic reticulum. Within 15 to 30 min of its formation, the proinsulin is transferred to the Golgi apparatus, where its conversion to insulin begins.[108] A combination of autoradiographic and immunocytochemical data have shown that proinsulin accumulates in a clathrin-coated Golgi-derived compartment in the presence of monensin.[109] This observation led the authors to the conclusion that the normal route for the processing of proinsulin is the passage into clathrin-coated membrane vesicles which later become mature secretory granules without clathrin. The transformation of proinsulin into insulin involves the removal of an intermediate peptide (the C peptide), which is stored in the secretory granules and is released along with insulin.[108] The cleavage is catalyzed by a thiol protease similar to that which converts proglucagon to glucagon.[110,111] In vitro, proteolysis can be inhibited by incubating the pancreatic B cells with analogs of lysine and arginine; however, the rate of secretion is unaffected under these conditions.[112] This result demonstrates that the discharge of the granules is independent of the nature of their contents.

Parathormone (PTH) is formed as preproPTH and thereafter cleaved first to proPTH and then to PTH.[113] The sequence of proPTH has been determined both by conventional techniques (Edman degradation) and by sequencing the cDNA corresponding to the RNA message. The tripeptide Lys-Lys-Arg occurs at the cleavage site. The proteolysis is effected by a trypsin-like protease; however classical trypsin inhibitors have no effect on this activity, although it is sensitive to chloroquine, an inhibitor of acid proteases such as cathepsin B. Information gathered from the timing of the intracellular transport of the polypeptide and the effects of inhibitors of events taking place in the RER and Golgi regions[114] suggest that the conversion of proPTH to PTH must be initiated in the Golgi apparatus but, once started, can continue in the secretory granules.

Another interesting example is the processing of proopiomelanocortin (POMC). This precursor gives rise to adrenocorticotropic hormone (ACTH) and β-lipotropin (LPH). This conversion takes place in the Golgi apparatus.[115] The LPH may be further processed in the secretory granules to yield β-endorphin. Similarly, ACTH may be converted to α-melanotropin (α-MSH) and corticotropin-like intermediate lobe peptide (CLIP) in the secretory granules; the carboxyl terminal of α-MSH is amidated. This α-amidation has been extensively studied. The activity is situated in the secretory granules and requires reduced ascorbate, copper (Cu^{2+}) ions and molecular oxygen. The peptide sequence recognized by the enzyme is X-gly, where X is amidated with the loss of the glycine residue. The name "peptidyl-glycine-amidating monoxygenase" has been suggested for the enzyme.[116,117] POMC is an example of a prohormone precursor which can give rise to several different hormonal activities as a result of successive proteolysis in various cellular compartments.

Procalcitonin is another precursor which is subject to two different proteolytic conversions, one taking place in the smooth endoplasmic reticulum and the other in the secretory granules.[118]

The maturation of proalbumin to albumin has been shown to begin in the *trans* cisternae of the Golgi apparatus from the results of experiments performed with drugs modifying the transfer from the rough endoplasmic reticulum to the Golgi, and with monensin.[119,120] An interesting point is that the inhibition of this conversion, brought about by monensin, does not prevent the secretion of proalbumin, although this is retarded. This parallels the situation with insulin cited above.

These examples illustrate the type of questions which can be posed with regard to the final post-translational modifications undergone by proteins and open up an enormous field of research, especially in molecular and cellular endocrinology.

3. Concentration of Secretory Products and Condensing Vacuoles

The reader will find that the process of concentration of secretory products has been described in several reviews.[48,86] In many cell types, this step occurs in the peripheral dilatations of the Golgi apparatus; in others, for example exocrine cells of the pancreas and parotid gland, it takes place in specialized condensing vacuoles. Usually, condensing vacuoles appear light and mature secretory granules dark upon electron microscopy. Until recently, this distinction could not be observed in goblet cells; however, after fixation of sections by quick freezing, a difference between immature and mature granules can be seen.[121]

The degree of concentration experienced by the products is often considerable but the mechanisms involved have not yet been elucidated. The process does not require an input of energy.[8,9,122] Several hypotheses have been put forward, such as the formation of aggregates between the secretory products and other components, for example sulfated polyanions[123,124] or calcium[125] As described above, the intermediary step between the *trans* cisternae of the Golgi apparatus and the mature secretory granules is also the point at which various modifications of protein structure occur. Thus, it is during this stage that the secretory products in the granules acquire the final form in which they are released outside the cell. Exocytosis is the second energy-dependent process.

In this first part, we have given an account of the present state of knowledge of the biosynthesis and post-translational modifications of secreted proteins. We will now attempt, briefly, to show how this information can provide some of answers to the questions posed in our introduction.

III. INTRACELLULAR TRANSPORT OF PROTEINS

In this section we will deal briefly with four major aspects of intracellular transport of proteins.

1. How are the proteins synthesized in the rough endoplasmic reticulum directed to their ultimate destination; that is, incorporated into various types of membranes or released into the lumen of the reticulum to be subsequently exported or included in lysosomes?
2. What is the role of the cytoskeleton in the intracellular transport of proteins?
3. Can the two possible pathways of secretion, constitutive and regulated, co-exist in the same cell? Are these the signals that direct proteins towards one route or the other?
4. Is the processing of proteins regulated?

A. Sorting of Proteins towards Membranes, Lysosomes, or Exportation

This subject has been reviewed several times.[15,85,126] In the previous section we described the processing of proteins destined for export during their passage through the various cellular compartments. Here we will consider how proteins are inserted into membranes or directed to the lysosomes.

1. Insertion of Proteins into Membranes

Many different membrane proteins have been studied, and we will mention just two examples, the membrane glycoproteins of enveloped viruses and immunoglobulins. Like secreted proteins, membrane proteins are synthesized with a "signal peptide" which initiates their passage through the reticulum membrane; but, in addition, they possess "halt" or "stop transfer" signals which seem to interrupt their translocation through the membrane and position the mature protein within the lipid bilayer.[15]

The example of secreted and membrane immunoglobulins is particularly interesting.[15,126] Depending on their functional state, lymphocytes synthesize immunoglobulins either for incorporation into the membrane or for secretion. These two types of immunoglobulin differ in the C-terminal peptides of their μ chains.[127] In the case of secreted immunoglobulin this peptide contains 20 amino acids and is mainly hydrophilic in character; in contrast, the C-terminal peptide of membrane immunoglobulin contains 41 residues and is hydrophobic. These two polypeptides are coded for by two different RNA messages.[128] The activation of the B cell by antigen causes a change in the transcription of RNA.[129] Thus, regulation at the transcriptional level occurs according to the functional state of the cell, and the immunoglobulins synthesized possess either a hydrophobic segment allowing insertion into the membrane, or a hydrophilic region which marks them out for secretion.

Another model which has been extensively studied is that of the membrane glycoproteins of enveloped viruses: in particular the hemagglutinin (HA) of the influenza virus and the G protein of vesicular stomatitis virus (VSV).[130-132] These proteins accumulate in large quantities in the membranes of infected cells, where they adopt a transmembrane orientation. In epithelial cells these two proteins are partitioned differently between the two poles of the cell. The HA glycoprotein is delivered to the apical surface, whereas the G protein is incorporated mainly in the basolateral region of the plasma membrane. These viral proteins may be used to study the recycling of membrane components. The G protein, when inserted artificially into the apical region, can be recycled and transported to the basolateral area.

Some very interesting experiments have been carried out by creating "chimeric genes", which can transform a protein destined for export into a membrane protein.[132] The hybrid polypeptides so formed consist of a secreted protein with the C-terminal portion of the HA or G glycoprotein spliced to its C-terminal. These hybrids inserted into the reticulum membrane are retained in the cell. It can also be recycled, as for the G protein. Therefore the C-terminal portion grafted onto the polypeptide must contain the "halt transfer" signal.

These few examples demonstrate that the insertion of a protein into the plasma membrane depends on a signal in the polypeptide sequence, and, as illustrated with reference to immunoglobulins, the presence or absence of this signal determines whether the proteins will be located in the membrane or secreted.

2. Transport of Lysosomal Proteins

The sorting of lysosomal proteins is the only known example of a process where the "addressing signal" is added at the level of the Golgi apparatus.[85] Lysosomal enzymes follow the same route as secreted proteins, and in the *cis* cisternae they are phosphorylated on position 6 of a mannose residue, as described in the previous section. The mannose-6-phosphate residues are recognized by a receptor while still in the Golgi complex, and this signal diverts the lysosomal enzymes from the secretory pathway. A mannose-6-phosphate receptor, of molecular weight 215 kdaltons, has been well characterized. Its location has been extensively debated; however, it has been shown to be present in the Golgi cisternae. This observation suggests that it is in this organelle that proteins are directed towards the lysosomes although other possibilities cannot be excluded.[85,133-135] Recently, a second mannose-6-phosphate receptor, of 46 kdaltons, which differs from the receptor described above in its requirement for manganese ions for the binding of the ligand, has been identified.[136] At present, the respective roles of these two receptors are not known.

In conclusion, it is clear that the destination of a protein synthesized in the rough endoplasmic reticulum may be determined either by the presence of a particular amino acid sequences or by a post-translational modification.

B. The Cytoskeleton and Intracellular Transport of Proteins

A considerable amount of work has been devoted towards understanding the role of the cytoskeleton, and particularly of microtubules, in the secretory process and the intracellular transport of proteins.[137-146] Most of these studies have made use of drugs which either disrupt the microtubule network, such as colchicine, vinblastine, and nocodazole, or those which stabilize it, like taxol. It was soon apparent that colchicine inhibited the release of newly synthesized proteins without affecting the induced discharge of proteins stored in secretory granules. When considering the results of all these studies it is important to distinguish between the short term and long-term effects of these drugs. In pulse-chase experiments followed by autoradiography performed in the rat lacrimal gland model, we were able to show that under conditions where the microtubules were destroyed, colchicine added before the radioactive pulse reduced the transfer of labeled proteins from the rough endoplasmic reticulum (RER) to the Golgi apparatus. In these conditions the transitory accumulation of proteins in the Golgi area was abolished before any structural alteration in this organelle was visible. In addition, colchicine inhibited the liberation of proteins from secretory granules which had been formed in the presence of the drug while having little effect on the discharge of secretory granules formed in its absence. Taking all these results together, we concluded that an intact microtubule system was not indispensible for the transport of proteins in the RER-Golgi area, but could facilitate this process by maintaining the spatial organization of this region. Colchicine might also alter the formation of secretory granules, rendering them incapable of fusing correctly with the plasma membrane.[139,140,142]

Recently, other workers have studied this problem by applying immunofluorescence techniques to follow the glycosylation and transport of the G protein of VSV.[144-146] They showed that changes in the state of assembly and distribution of microtubules did not significantly affect the transfer of the G-protein from the RER to the Golgi apparatus or to the plasma membrane, or the incorporation of sialic acid into the glycoprotein. It had already been shown that colchicine had no effect on the production of VSV particles by cultured cells.[147] However, under their experimental conditions, the authors noted that colchicine modified the configuration and localization of the elements of the Golgi complex.[146] They reported that although the rate of appearance of the G protein at the cell surface was unchanged, the polarity of insertion into the membrane was altered.[146]

From the observations summarized above it appears that the microtubule system does not play a crucial role in the intracellular transport of secreted and membrane proteins, but may be essential to maintain the necessary intracellular architecture for polarized transport to particular regions of the plasma membrane.

C. Constitutive and Regulated Secretion

In certain cell types, for example fibroblasts, chondrocytes, macrophages, lymphocytes, and some cell lines, a pathway for nonregulated secretion is found as we described in Section II. In the process of regulated secretion the products may be considerably concentrated[148] then stored in granules and discharged in response to an external stimulus. In the constitutive pathway the products are continuously released by exocytosis. This subject has been recently reviewed.[149]

Some extremely interesting experiments have been performed by Kelly and co-workers[150-152] concerning the relationship between these two routes of secretion. They used the AtT20 cell line, which produces proopiomelanocortin (POMC), the precursor of ACTH and endorphins (see Section II). ACTH is stored in dense granules and can be released in response to an external stimulus. Besides this regulated secretory activity, these cells also have the ability to release POMC and laminin constitutively. By transferring genes coding for either secreted or membrane proteins into these cells it was possible to obtain an insight into the sorting of proteins between the two pathways of secretion. When a protein which is normally released in a regulated fashion is introduced into the cells, this protein follows the regulated pathway. On the other hand, when a membrane protein, such as a modified form of the G protein from VSV is involved, it takes the constitutive route. Furthermore, proinsulin, introduced by genetic transfer, is converted to insulin, as POMC is converted to ACTH. Therefore, a signal, as yet unknown, must be present in the amino acid sequence of proteins and serves to direct them towards one or the other route. These experimental results also show that the two pathways of secretion can co-exist in the same cell, but that a particular protein is dispatched towards only one of the two possibilities.

D. Regulation of the Processing of Proteins

We have already referred to regulation at the transcriptional level with respect to the synthesis of two different immunoglobulin messages at different times by lymphocytes. We will now consider the regulation of the N-glycosylation of proteins.

Thyroid stimulating hormone (TSH) increases the synthesis of thyroglobulin and stimulates the activity of several glycosyl transferases, in particular the oligosaccharide transferase responsible for the transfer of the oligosaccharide to an asparagine residue of the nascent chain.[153] In the hen oviduct and bovine pancreas, dolicholphosphate has been shown to be the rate-limiting factor for glycosylation.[154] In hepatic microsomes from rats pretreated with an inflammatory agent (turpentine) the incorporation of tritiated mannose into dichol-phosphate is doubled compared to the control, probably due to an accumulation of this agent in the membrane fraction. On the other hand, treatment with dexamethasone leads to an increase in mannosyl transferase activity.[155]

In the parotid gland it has been shown that isoproterenol, which activates β-adrenergic receptors, increases the incorporation of tritiated mannose into N-linked glycoproteins, while only slightly affecting the incorporation of amino acids. This effect is probably dependent on cAMP, since it can be partially reproduced by forskolin, which activates the adenylcyclase directly.[156,157] Recently, it has been shown that β-adrenergic stimulation of the parotid gland can cause an increased synthesis of mannosylphosphodolichol and oligosaccharide-dolicholpyrophosphate, as well as increased turnover of the latter.[158]

These examples show that N-glycosylation can be regulated independently of the synthesis of the protein moiety, mainly at two levels: first by the concentration of dolicholphosphate and second by the activity of glycosyl transferases and the turnover of the oligosaccharide-dolicholpyrophosphate intermediate.

IV. CONCLUDING REMARKS

As this short review illustrates, the secretion and intracellular movement of proteins constitute a large and rich domain for present-day biochemistry and cell biology. While the major outlines and principal mechanisms are being established, there remains a large number of open questions which offer a wide field for research. Some of these questions are listed below.

If certain proteins acquire signals allowing them to be inserted into membranes or lysosomes, is there also a signal (with receptor) corresponding to exported proteins? How are secreted proteins directed to either the regulated or the constitutive pathway? What is the role of the cytoskeleton?

Is there a relationship between the coating and uncoating of vesicles with clathrin and the intracellular transport of proteins and its dependence on energy?

How is the concentration of products in the condensing vacuoles achieved?

What are the biochemical mechanisms involved in the regulation (via intracellular messengers such as cAMP) of the processing of proteins? This last question brings into focus the relationship between cell biology and pharmacology.

A multidisciplinary approach, invoking techniques drawn from biochemistry, molecular biology, and cell biology, in the areas which we have just cited (signals and receptors, regulatory mechanisms, the cytoskeleton) should lead to a better understanding of the secretory process in the new future.

REFERENCES

1. **Palade, G.,** Intracellular aspects of the process of protein synthesis, *Science,* 189, 347, 1975.
2. **Milstein, C., BrownLee, G. G., Harrison, T. M., and Mathews, M. B.,** A possible precursor of immunoglobulin light chains, *Nature (London), New Biol.,* 239, 117, 1972.
3. **Redman, C. M. Siekevitz, P., and Palade, G. E.,** Synthesis and transfer of amylase in pigeon pancreatic microsomes, *J. Biol. Chem.,* 241, 1150, 1966.
4. **Anfinsen, C. B.,** The formation of the tertiary structure of proteins, *Harvey Lect.,* 61, 95, 1966.
5. **Molnar, J., Robinson, G. B., and Winzler, R. J.,** Biosynthesis of glycoproteins. IV. The subcellular sites of incorporation of glucosamine 1 ¹⁴C into glycoproteins in rat liver, *J. Biol. Chem.,* 240, 1882, 1965.
6. **Olsen, B. R., Berg, R. A., Kishida, Y., and Prockop, D. J.,** Collagen synthesis: localisation of prolyl hydroxylase in tendon cells detected with ferritin-labeled antibodies, *Science,* 182, 825, 1973.
7. **Jamieson, J. D. and Palade, G. E.,** Intracellular transport of secretory proteins in the pancreatic exocrine cell. I. Role of the peripheral elements of the Golgi complex, *J. Cell. Biol.,* 34, 577, 1967.
8. **Jamieson, J. D. and Palade, G. E.,** Intracellular transport of secretory proteins in the pancreatic exocrine cell. IV. Metabolic requirements, *J. Cell. Biol.,* 39, 589, 1968.

9. **Jamieson, J. D. and Palade, G. E.,** Intracellular transport of secretory proteins in the pancreatic exocrine cell. II. Transport to condensing vacuoles and zymogen granules, *J. Cell. Biol.*, 34, 597, 1967.

10. **Tartakoff, A., Greene, L. J., and Palade, G. E.,** Studies on the guinea pig pancreas. Fractionation and partial characterization of exocrine proteins, *J. Biol. Chem.*, 249, 7420, 1974.

11. **Neutra, M. and Leblond, C. P.,** Radioautographic comparison of the uptake of galactose-^{3}H and glucose-^{3}H in the Golgi region of various cells secreting glycoproteins or mucopolysaccharides, *J. Cell. Biol.*, 30, 1966, 137.

12. **Young, R. W.,** The role of the Golgi complex in sulfate metabolism, *J. Cell. Biol.*, 57, 175, 1973.

13. **Gething, M. J.,** Introduction, in *Protein Transport and Secretion, Current Communications in Molecular Biology*, Gething, M. J., Ed., Cold Spring Harbor Laboratory, Cold Spring Harbor, New York, 1985, 1.

14. **Kreil, G.,** Transfer of proteins across membranes, *Ann. Rev. Biochem.*, 50, 317, 1981.

15. **Sabatini, D. D., Kreibich, G., Morinoto, T., and Adenisk, M.,** Mechanisms for the incorporation of proteins in membranes and organelles, *J. Cell. Biol.*, 92, 1, 1982.

16. **Walter, P., Gilmore, R., and Blobel, G.,** Protein translocation across the endoplasmic reticulum, *Cell*, 38, 5, 1984.

17. **Gilmore, R., Walter, P., Anderson, D., Erickson, A., Muller, M., and Blobel, G.,** The mechanism of protein translocation across the endoplasmic reticulum membrane, in *Protein Transport and Secretion, UCLA Symposia on Molecular and Cellular Biology*, Vol. 15, Oxender, D. L., Ed., Alan R. Liss, New York, 1984, 291.

18. **Lingappa, V. R., Katz, F. N., Lodish, H. F., and Blobel, G.,** A signal sequence for the insertion of a transmembrane glycoprotein. Similarities to the signals of secretory proteins in primary structure and function, *J. Biol. Chem.*, 253, 8667, 1978.

19. **Erickson, A. H. and Blobel, G.,** Early events in the biosynthesis of the lysosomal enzyme cathepsine D, *J. Biol. Chem.*, 254, 11771, 1979.

20. **Borgese, D., Blobel, G., and Sabatini, D. D.,** *In vitro* exchange of ribosomal subunits between free and membrane-bound ribosomes, *J. Mol. Biol.*, 74, 415, 1973.

21. **Adelman, M. R., Blobel, G., and Sabatini, D. D.,** An improved cell fractionation procedure for the preparation of rat liver membrane bound ribosomes, *J. Cell. Biol.*, 56, 191, 1973.

22. **Adelman, M. R., Sabatini, D. D., and Blobel, G.,** Ribosome-membrane interaction. Nondestructive disassembly of rat liver rough microsomes into ribosomal and membranous components, *J. Cell. Biol.*, 56, 206, 1973.

23. **Sabatini, D. D. and Blobel, G.,** Controlled proteolysis of nascent polypeptides in rat liver cell fractions. II. Location of the polypeptides in rough microsomes, *J. Cell. Biol.*, 45, 146, 1970.

24. **Blobel, G. and Dobberstein, B.,** Transfer of proteins across membranes. I. Presence of proteolytically processed and unprocessed nascent immunoglobulin light chains on membrane-bound ribosomes of murine myeloma, *J. Cell. Biol.*, 67, 835, 1975.

25. **Blobel, G. and Dobberstein, B.,** Transfer of protein across membranes. II. Reconstitution on functional rough microsomes from heterologous components, *J. Cell. Biol.*, 67, 852, 1975.

26. **Devillers-Thiery, A., Kindt, T., Scheele, G., and Blobel, G.,** Homology in amino-terminal sequence of precursors to pancreatic secretory proteins, *Proc. Natl. Acad. Sci. U.S.A.*, 72, 5016, 1975.

27. **Kemper, B., Habener, J. F., Mulligan, R. C., Potts, J. T. Jr., and Rich, A.,** Preproparathyroid hormone: a direct translation product of a parathyroid messenger RNA, *Proc. Natl. Acad. Sci. U.S.A.*, 71, 3731, 1974.

28. **Meyer, D. I., Krause, E., and Dobberstein, B.,** Secretory protein translocation across membranes. The role of the ''docking protein'', *Nature (London)*, 297, 647, 1982.

29. **Gilmore, R., Blobel, G., and Walter, P.,** Protein translocation across the endoplasmic reticulum. I. Detection in the microsomal membrane of a receptor for the signal recognition particle, *J. Cell. Biol.*, 95, 463, 1982.

30. **Gilmore, R., Walter, P., and Blobel, G.,** Protein translocation across the endoplasmic reticulum. II. Isolation and characterization of the signal recognition particle receptor, *J. Cell. Biol.*, 95, 470, 1982.

31. **Walter, P. and Blobel, G.,** Signal recognition particle contains a 7 S RNA essential for protein translocation across the endoplasmic reticulum, *Nature (London)*, 299, 691, 1982.

32. **Walter, P., Ihrahimi, I., and Blobel, G.,** Translocation of protein across the endoplasmic reticulum. I. Signal recognition protein (SRP) binds to *in vitro*-assembled polysomes synthesizing secretory protein, *J. Cell. Biol.*, 91, 545, 1981.

33. **Walter, P. and Blobel, G.,** Translocation of proteins across the endoplasmic reticulum. II. Signal recognition protein (SRP) mediates the selective binding to microsomal membranes of *in vitro*-assembled polysomes synthesizing secretory protein, *J. Cell. Biol.*, 91, 551, 1981.

34. **Walter, P. and Blobel, G.,** Translocation of protein across the endoplasmic reticulum. III. Signal recognition protein (SRP) causes signal sequence-dependent and site-specific arrest of chain elongation that is released by microsomal membranes, *J. Cell. Biol.*, 91, 557, 1981.

35. **Walter, P. and Blobel, G.,** Subcellular distribution of signal recognition particle and 7 SL-RNA determined with polypeptide-specific antibodies and complementary DNA probe, *J. Cell. Biol.,* 97, 1693, 1983.
36. **Hortsch, M., Avossa, D., and Meyer, D. I.,** Factors mediating protein translocation in the endoplasmic reticulum: the docking protein and beyond, in *Protein Transport and Secretion, Current Communications in Molecular Biology,* Gething, M. J., Ed., Cold Spring Harbor Laboratory, Cold Spring Harbor, New York, 1985, 24.
37. **Kreibich, G., Ulrich, B. L., and Sabatini, D. D.,** Proteins of rough microsomal membranes related to ribosome binding. I. Identification of ribophorines I and II membrane proteins characteristic of rough microsomes, *J. Cell. Biol.,* 77, 464, 1978.
38. **Carne, T. and Scheele, G.,** Role of presecretory proteins in the secretory process, in *Cell Biology of the Secretory Process,* S. Karger, Basel, 1984, 73.
39. **Scheel, G. A.,** Biosynthesis, segregation, and secretion of exportable proteins by the exocrine pancreas, *Am. J. Physiol.,* 238, G467, 1980.
40. **Palmiter, R. D., Gagnon, J., and Walsh, K. A.,** Ovalbumin: a secreted protein without a transcient hydrophobic leader sequence, *Proc. Natl. Acad. Sci. U.S.A.,* 75, 94, 1978.
41. **Lingappa, V. R., Lingappa, J. R., and Blobel, G.,** Chicken ovalbumin contains an internal signal sequence, *Nature (London),* 281, 117, 1979.
42. **Davis, B. D. and Tai, P. C.,** The mechanism of protein secretion across membranes, *Nature (London),* 283, 433, 1980.
43. **Borgese, N., Mok, W., Kreilich, G., and Sabatini, D. D.,** Ribosome-membrane interaction: *in vitro* binding of ribosomes to microsomal membranes, *J. Mol. Biol.,* 88, 559, 1974.
44. **Von Heijne, G. and Blomberg, C.,** Trans-membrane translocation of proteins (the direct transfer model), *Eur. J. Biochem.,* 97, 175, 1979.
45. **Von Heinjne, G.,** Structural and thermodynamic aspects of the transfer of proteins into and across membranes, in *Membrane Protein Biosynthesis and Turnover,* Vol. 24, *Current Topics in Membranes and Transport,* Bronner, F., Ed., Academic Press, New York, 1985, 151.
46. **Clauser, H., Herman, G., Rossignol, B., and Harbon, S.,** Biosynthesis of glycoproteins. Biosynthesis at the cellular and subcellular level, in *Glycoproteins,* Part B, Gottschalk, A., Ed., Elsevier, Amsterdam, 1972, 1151.
47. **Kornfeld, R. and Kornfeld, S.,** Assembly of asparagine-linked oligosaccharides, *Ann. Rev. Biochem.,* 54, 631, 1985.
48. **Tartakoff, A. M.,** The confined function — model of the Golgi complex center for ordered processing of biosynthetic products of rough endoplasmic reticulum, *Int. Rev. Cytol.,* 85, 221, 1983.
49. **Robbins, P. W.,** Protein glycosylation in molecular biology, in *Protein Transport and Secretion, Current Communications in Molecular Biology,* Gething, M. J., Ed., Cold Spring Harbor Laboratory, Cold Spring Harbor, N.Y., 1985, 109.
50. **Snider, M. D. and Rogers, O. C.,** Transmembrane movement of oligosaccharide lipids during glycoprotein synthesis, *Cell,* 36, 753, 1984.
51. **Snider, M. D. and Robbins, P. W.,** Transmembrane organization of protein glycosylation. Mature oligosaccharide-lipid is located on the luminal side of microsomes from Chinese hamster ovary cells, *J. Biol. Chem.,* 257, 6796, 1982.
52. **Hanover, J. A. and Lennarz, W. J.,** Transmembrane assembly of N-linked glycoproteins. Studies on the topology of saccharide-lipid synthesis, *J. Biol. Chem.,* 257, 2787, 1982.
53. **Haselbeck, A. and Tanner, W.,** Dolichyl phosphate mediated mannosyl transfer through liposomal membranes, *Proc. Natl. Acad. Sci. U.S.A.,* 79, 1520, 1982.
54. **Turco, S. J.,** Modification of oligosaccharide-lipid synthesis and protein glycosylation in glucose deprived cells, *Arch. Biochem. Biophys.,* 205, 330, 1980.
55. **Gershman, H. and Robbins, P. W.,** Transitory effects of glucose starvation on the synthesis of dolichol linked oligosaccharides in mammalian cells, *J. Biol. Chem.,* 256, 7774, 1981.
56. **Rearick, J. I., Chapman, A., and Kornfeld, S.,** Glucose starvation alters lipid-linked oligosaccharide biosynthesis in Chinese hamster ovary cells, *J. Biol. Chem.,* 256, 6255, 1981.
57. **Spiro, R. G., Spiro, M. J., and Bhoyroo, V. D.,** Studies on the regulation of the biosynthesis of glucose-containing oligosaccharide-lipids. Effect of energy deprivation, *J. Biol. Chem.,* 258, 9469, 1983.
58. **Hart, G. W., Brew, K., Grant, G. A., Bradshaw, R. A., and Lennarz, W. J.,** Primary structural requirements for the enzymatic formation of the N-glycosidic bond in glycoproteins. Studies with natural and synthetic peptides, *J. Biol. Chem.,* 254, 9747, 1979.
59. **Bause, E. and Legler, G.,** The role of the hydroxy amino-acid in the triplet-sequence Asn-Xaa-Thr (Ser) for the N-glycosylation step during glycoprotein biosynthesis, *Biochem. J.,* 195, 639, 1981.
60. **Pless, D. D. and Lennarz, W. J.,** Enzymatic conversion of proteins to glycoproteins, *Proc. Natl. Acad. Sci. U.S.A.,* 74, 134, 1977.
61. **Atkinson, P. H. and Lee, J. T.,** Co-translational excision of α-glucose and α-mannose in nascent vesicular stomatitis virus G protein, *J. Cell. Biol.,* 98, 2245, 1984.

62. **Hubbard, S. C. and Robbins, P. W.,** Synthesis and processing of protein-linked oligosaccharides in vivo, *J. Biol. Chem.,* 254, 4568, 1979.

63. **Hay, E. D.,** Extracellular matrix, *J. Cell. Biol.,* 91, 205S, 1981.

64. **Fessler, J. H. and Fessler, L. I.,** Biosynthesis of procollagen, *Ann. Rev. Biochem.,* 47, 129, 1978.

65. **Lindhal, U. and Höök, M.,** Glycosaminoglycans and their binding to biological macromolecules, *Ann. Rev. Biochem.,* 47, 385, 1978.

66. **Diegelmann, R. F., Bernstein, L., and Peterkofsky, B.,** Cell-free collagen synthesis on membrane bound polysomes of chick embryo-connective tissue and the localization of prolyl hydroxylase on the polysome-membrane complex, *J. Biol. Chem.,* 248, 6514, 1973.

67. **Housley, T. J., Rowland, F. N., Ledger, P. W., Kaplan, J., and Tanzer, M. L.,** Effects of tunicamycin on the biosynthesis of procollagen by human fibroblasts, *J. Biol. Chem.,* 255, 121, 1980.

68. **Olsen, B. R., Berg, R. A., Kishida, Y., and Prockop, D. J.,** Further characterization of embryonic tendon fibroblasts and the use of immunoferritin techniques to study collagen biosynthesis, *J. Cell. Biol.,* 64, 340, 1975.

69. **Cardinal, G. J. and Udenfriend, S.,** Prolyl hydroxylase, *Adv. Enzymol,* 41, 245, 1974.

70. **Wold, F.,** *in vivo* chemical modification of proteins (post-translational modification), *Ann Rev. Biochem.,* 50, 783, 1981.

71. **Freedman, R. B.,** Native disulphide bond formation in protein biosynthesis: evidence for the role of protein disulphide isomerase, *Trends Biochem. Sci.,* 9, 438, 1984.

72. **Scheel, G. and Jacoby, R.,** Conformational changes associated with proteolytic processing of presecretory proteins allow glutathione catalyzed formation of native disulfide bonds, *J. Biol. Chem.,* 257, 12277, 1982.

73. **Roth, R. A. and Koshland, M. E.,** Role of disulfide interchange enzyme in immunoglobulin synthesis, *Biochemistry,* 20, 6594, 1981.

74. **Brockway, B. E., Forster, S. J., and Freedman, R. B.,** Protein disulphide-isomerase activity in chick-embryo tissues. Correlation with the biosynthesis of procollagen, *Biochem. J.,* 191, 873, 1980.

75. **Lambert, N. and Freedman, R. B.,** Structural properties of homogeneous protein disulphide-isomerase from bovine liver purified by a rapid high-yielding procedure, *Biochem. J.,* 213, 225, 1983.

76. **Ohba, H., Harano, T., and Omura, T.,** Intracellular and intramembranous localization of a protein disulfide isomerase in rat liver, *J. Biochem.,* 89, 889, 1981.

77. **Mills, E. N. C., Lambert, N., and Freedman, R. B.,** Identification of protein disulphide-isomerase as a major acidic polypeptide in rat liver microsomal membranes, *Biochem. J.,* 213, 245, 1983.

78. **Hortin, G. and Boime, I.,** Inhibition of preprotein processing in ascites tumor lysates by incorporation of a leucine analog, *Proc. Natl. Acad. Sci. U.S.A.,* 77, 1356, 1980.

79. **Gordon, J. I., Smith, D. P., Andy, R., Alpers, D. H., Schonfeld, G., and Strauss, A. W.,** The primary translation product of rat intestinal apolipoprotein A-I mRNA is an unusual preproprotein, *J. Biol. Chem.,* 257, 971, 1982.

80. **Hortin, G. and Boime, I.,** Transport of an uncleaved preprotein into the endoplasmic reticulum of rat pituitary cells, *J. Biol. Chem.,* 256, 1491, 1981.

81. **Hortin, G. and Boime, I.,** Markers for processing sites in eukaryotic proteins: characterization with amino acid analogs, *Trends Biochem. Sci.,* 8, 320, 1983.

82. **Schwarz, R. T. and Datema, R.,** Inhibitors of trimming: new tools in glycoprotein research, *Trends Biochem. Sci.,* 9, 32, 1984.

83. **Elbein, A. D.,** Inhibitors of glycoprotein synthesis, *Meth. Enzymol. Biomembranes,* 98, 135, 1983.

84. **Farquhar, M. G.,** Intracellular membrane traffic: Pathways, carriers, and sorting devices, *Meth. Enzymol. Biomembranes,* 98, 1, 1983.

85. **Farquhar, M. G.,** Progress in unraveling pathways of Golgi traffic, *Annu. Rev. Cell. Biol.,* 1, 447, 1985.

86. **Farquhar, M. G. and Palade, G. E.,** The Golgi apparatus (complex) — (1954—1981) — from artifact to center stage, *J. Cell. Biol.,* 91, 77s, 1981.

87. **Novikoff, A. B.,** The endoplasmic reticulum: a cytochemist's view (a review), *Proc. Natl. Acad. Sci. U.S.A.,* 73, 2781, 1976.

88. **Roth, J. and Berger, E. G.,** Immunocytochemical localization of galactosyltransferase in Hela cells: codistribution with thiamine pyrophosphatase in trans-Golgi cisternae, *J. Cell. Biol.,* 92, 223, 1982.

89. **Smith, C. E.,** Ultrastructural localization of nicotinamide adenine dinucleotide phosphatase (NADPase) activity to the intermediate saccules of the Golgi apparatus in rat incisor ameloblasts, *J. Histochem. Cytochem.,* 28, 16, 1980.

90. **Dunphy, W. G., Brands, R., and Rothman, J. E.,** Attachment of terminal N-acetylglucosamine to asparagine-linked oligosaccharides occurs in central cisternae of the Golgi stack, *Cell,* 40, 463, 1985.

91. **Friend, D. S. and Murray, M. J.,** Osmium impregnation of the Golgi apparatus, *Am. J. Anat.,* 117, 135, 1965.

92. **Morre, D. J. and Ovtracht, L.,** Dynamics of the Golgi apparatus: membrane differentiation and membrane flow, *Int. Rev. Cytol.,* 5, 61, 1977.

93. **Rothman, J. E.,** The Golgi apparatus: two organelles in tandem, *Sciences*, 213, 1212, 1981.
94. **Tartakoff, A. M.,** Perturbation of the structure and function of the Golgi complex by monovalent carboxylic ionophore, in *Meth. Enzymol. Biomembranes*, 98, 47, 1983.
95. **Roth, J., Lucocq, J. M., Berger, E. C., Paulson, J. C., and Watkins, W. M.,** Terminal glycosylation is compartmentalized in the Golgi apparatus, *J. Cell. Biol.*, 99, 229a, 1984.
96. **Dunphy, W. G., Brands, R., and Rothman, J. E.,** Attachment of terminal N-acetylglucosamine to asparagine-linked oligosaccharides occurs in central cisternae of the Golgi stack, *Cell*, 40, 463, 1985.
97. **Novikoff, P. M., Tulsiani, D. R. P., Touster, O., Yann, A., and Novikoff, A. B.,** Immunocytochemical localization of α-D-mannosidase II in the Golgi apparatus of rat liver, *Proc. Natl. Acad. Sci. U.S.A.*, 80, 4364, 1983.
98. **Capasso, J. M. and Hirschberg, C. B.,** Mechanisms of glycosylation and sulfation in the Golgi apparatus: evidence for nucleotide sugar/nucleoside monophosphate and nucleotide sulfate/nucleoside monophosphate antiports in the Golgi apparatus membrane, *Proc. Natl. Acad. Sci. U.S.A.*, 81, 7051, 1984.
99. **Cunnings, R. D., Kornfeld, S., Schneider, W., Hobgood, K. K., and Tolleshauge, M.,** Biosynthesis of N- and O-linked oligosaccharides of the low density lipoprotein receptor, *J. Biol. Chem.*, 258, 15261, 1983.
100. **Kimura, J. H., Lohmander, L. S., and Hascall, V. C.,** Studies on the biosynthesis of cartilage proteoglycan in a model system of cultured chondrocytes from the Swarm rat chondrosarcoma, *J. Cell. Biochem.*, 26, 261, 1984.
101. **Heifetz, A., Watson, C., Johnson, A. R., and Roberts, M. K.,** Sulfated glycoproteins secreted by human vascular endothelial cells, *J. Biol. Chem.*, 257, 13581, 1982.
102. **Tajiri, K., Uchida, N., and Tanzer, M. L.,** Undersulfated proteoglycans are secreted by cultured chondrocytes in the presence of the ionophore monensin, *J. Biol. Chem.*, 255, 6036, 1980.
103. **Reitman, M. L. and Kornfeld, S.,** U.D.P.-N-acetylglucosamine: glycoprotein N-acetylglucosamine-1-phosphotransferase, *J. Biol. Chem.*, 256, 4275, 1981.
104. **Goldberg, D. E. and Kornfeld, S.,** Evidence for extensive subcellular organization of asparagine linked oligosaccharide processing and lysosomal enzyme phosphorylation, *J. Biol. Chem.*, 258, 3159, 1983.
105. **Docherty, K. and Steiner, D.,** Post-translational proteolysis in polypeptide hormone biosynthesis, *Annu. Rev. Physiol.*, 44, 625, 1982.
106. **Zimmerman, M., Mumford, R. A., and Steiner, D. F.,** Precursor processing in the biosynthesis of proteins, *Ann. N.Y. Acad. Sci.*, Vol. 343, 1980.
107. **Scott, J., Patterson, S., Rall, L., Bell, G. I., Crowford, R., Penschow, J., Niall, H., and Coghlam, J.,** The structure and biosynthesis of epidermal growth factor precursor, *J. Cell.*, Suppl. 3, 19, 1985.
108. **Gold, G. and Grodsky, G. M.,** The secretory process in B cells of the pancreas, in *Cell Biology of the Secretory Process*, Cantin, M., Ed., S. Karger, Basel, 1984, 359.
109. **Orci, L., Halban, P., Amherdt, M., Ravazzola, M., Vassalli, J. D., and Perrelet, A.,** A clathrin-coated, Golgi-related compartment of the insulin secreting cell accumulates proinsulin in the presence of monensin, *Cell*, 39, 39, 1984.
110. **Fletcher, D. J., Quigley, J. P., Bauer, E., and Noc, B. D.,** Characterization of proinsulin- and proglucagon-converting activities in isolated islet secretory granules, *J. Cell. Biol.*, 90, 312, 1981.
111. **Docherty, K., Carroll, R. J., and Steiner, D. F.,** Conversion of proinsulin to insulin: involvement of a 31,500 molecular weight thiol protease, *Proc. Natl. Acad. Sci. U.S.A.*, 79, 4613, 1982.
112. **Halban, P. A.,** Inhibition of proinsulin to insulin conversion in rat islets using arginine and lysine analogs. Lack of effect on rate of release of modified products, *J. Biol. Chem.*, 257, 13177, 1982.
113. **Kemper, B.,** Biosynthesis and secretion of parathyroid hormones, in *Cell Biology of the Secretory Process*, Cantin, M., Ed., S. Karger, Basel, 1984, 443.
114. **Chu, L. L. H., McGregor, R. R., and Cohn, D. V.,** Energy-dependent intracellular translocation of proparathormone, *J. Cell. Biol.*, 72, 1, 1977.
115. **Glembotski, C. C.,** Subcellular fractionation studies on the post translational processing of proadrenocorticotropic hormone/endorphine in rat intermediate pituitary, *J. Biol. Chem.*, 256, 7433, 1981.
116. **Glembotski, C. C.,** The α-amidation of α-melanocyte stimulating hormone in intermediate pituitary requires ascorbic acid, *J. Biol. Chem.*, 259, 13041, 1984.
117. **Mains, R. E., Glembotski, C. C., and Eipper, B. A.,** Peptide α-amidation activity in mouse anterior pituitary At T-20 cell granules: properties and secretion, *Endocrinology*, 114, 1522, 1984.
118. **Treilhou-Lahille, F.,** personnal communication.
119. **Redman, C. M., Banerjee, D., Manning, C., Huang, C. Y., and Green, K.,** *In vivo* effect of colchicine on hepatic protein synthesis and on the conversion of proalbumin to serum albumin, *J. Cell. Biol.*, 77, 400, 1978.
120. **Mizumi, Y., Oda, K., Takami, N., and Ikehara, Y.,** Accumulation and delayed secretion of unprocessed proteins by monensin in cultured rat hepatocytes, in *Protein Transport and Secretion, UCLA Symposia on Molecular and Cellular Biology*, Vol. 15, Oxender, D. L., Ed., Alan, R. Liss, New York, 1984, 365.

121. **Sandoz, D., Nicolas, G., and Laine, M. C.,** Two mucous cell types revisited after quick-freezing and cryosubstitution, *Biol. Cell.*, 54, 79, 1985.
122. **Jamieson, J. D. and Palade, G. E.,** Condensing vacuole conversion and zymogen granule discharge in pancreatic exocrine cells: metabolic studies, *J. Cell. Biol.*, 48, 503, 1971.
123. **Reggio, H. A. and Palade, G. E.,** Sulfated compounds in the zymogen granules of the guinea pig pancreas, *J. Cell. Biol.*, 77, 288, 1978.
124. **Giamattasio, G., Zanini, A., Rosa, P., Meldolesi, J., Margolis, R. K., and Margolis, R. U.,** Molecular organization of prolactin granules. III. Intracellular transport of sulfated glycosaminoglycans and glycoproteins of the bovine prolactin granule matrix, *J. Cell. Biol.*, 86, 273, 1980.
125. **Clemente, F. and Meldolesi, J.,** Calcium and pancreatic secretion. Subcellular distribution of calcium and magnesium in the exocrine pancreas of the guinea pig, *J. Cell. Biol.*, 65, 88, 1975.
126. **Rodriduez-Boulan, E., Misek, D. E., Vega de Salas, D., Salas, P. J. I., and Bard, E.,** Protein sorting in the secretory pathway in membrane protein biosynthesis and turnover, in *Current Topics in Membranes and Transport*, Bronner, F. Ed., Academic Press, New York, 1985, 251.
127. **Kehry, M., Ewald, S., Douglas, R., Sibley, C., Raschke, W., Fambrough, D., and Hood, L.,** The immunoglobulin μ-chains of membrane-bound and secreted IgM molecules differ in their C-terminal segments, *Cell*, 21, 393, 1980.
128. **Rogers, J., Early, P., Carter, C., Calame, K., Bond, M., Hood, L., and Wall, R.,** Two mRNAs with different 3′ ends encode membrane-bound and secreted forms of immunoglobulin μ-chains, *Cell*, 20, 303, 1980.
129. **Early, P., Rogers, J., Davis, M., Calame, K., Bond, M., Wall, R., and Hood, L.,** Two mRNAs can be produced from a single immunoglobulin μ-gene by alternative RNA processing pathways, *Cell*, 20, 313, 1980.
130. **Matlin, K. S. and Simons, K.,** Sorting of an apical plasma membrane protein before it reaches the cell surface in cultured epithelial cells, *J. Cell. Biol.*, 99, 2131, 1984.
131. **Rindler, M. J., Ivanov, I. E., Plesken, H., and Sabatini, D. D.,** Polarized delivery of viral glycoproteins to the apical and basolateral plasma membranes of Madin-Darby canine kidney cells infected with temperature sensitive viruses, *J. Cell. Biol.*, 100, 136, 1985.
132. **Rizzolo, L. J., Gonzalez, A., Gottlieb, T. A., Finidori, J., Ivanov, I. E., Rindler, M. J., Adesnik, M. B., and Sabatini, D. D.,** Intracellular sorting and distinct recycling patterns of viral glycoproteins in polarized epithelial cells, in *Protein Transport and Secretion*, Current Communications in Molecular Biology, Gething, M. J., Ed., Cold Spring Harbor Laboratory, Cold Spring Harbor, New York, 1985, 147.
133. **Brown, W. J. and Farquhar, M. G.,** The mannose-6-P receptor for lysosomal enzymes is concentrated in cis Golgi cisternae, *Cell*, 36, 295, 1984.
134. **Geuze, H. J., Slot, J. W., Strous, G. J. A. M., Hasilik, A., and Figura, K. V.,** Ultrastructural localization of the mannose-6-phosphate receptor in rat liver, *J. Cell. Biol.*, 98, 2047, 1984.
135. **Gabel, C. A., Goldberg, D. E., and Kornfeld, S.,** Identification and characterization of cells deficient in the mannose-6-phosphate receptor: evidence for an alternate pathway for lysosomal enzyme targeting, *Proc. Natl. Acad. Sci. U.S.A.*, 80, 775, 1983.
136. **Hoflack, B. and Kornfeld, S.,** Lysosomal enzyme binding to mouse P388 D₁ macrophage membranes lacking the 215 Kd mannose-6-phosphate receptor: evidence for the existence of a second mannose-6-phosphate receptor, *Proc. Natl. Acad. Sci. U.S.A.*, 80, 4428, 1985.
137. **Reaven, E. P. and Reaven, G. M.,** Evidence that microtubules play a permissive role in hepatocyte very low density lipoprotein secretion, *J. Cell. Biol.*, 84, 28, 1980.
138. **Rossignol, B., Herman, G., and Keryer, G.,** Inhibition by colchicine of carbamylcholine induced glycoprotein secretion by the submaxillary gland. A possible mechanism of cholinergic induced protein secretion, *FEBS Lett.*, 21, 189, 1972.
139. **Chambaut-Guérin, A. M., Muller, P., and Rossignol, B.,** Microtubules and protein secretion in rat lacrimal glands. Relationship between colchicine binding and its inhibitory effect on the intracellular transport of proteins, *J. Biol. Chem.*, 253, 3870, 1978.
140. **Busson-Mabillot, S., Chambaut-Guérin, A. M., Ovtracht, L., Muller, P., and Rossignol, B.,** Microtubules and protein secretion in rat lacrimal glands: localization of short-term effects of colchicine on the secretory process, *J. Cell. Biol.*, 95, 105, 1982.
141. **Redman, C. M.,** Role of cytoskeleton in liver: *in vivo* effect of colchicine on hepatic protein secretion, *Meth. Enzymol.*, 98, 169, 1983.
142. **Rossignol, B., Chambaut-Guérin, A. M., and Muller, P.,** Role of cytoskeleton in secretory processes: lacrimal and salivary glands, in *Meth. Enzymol.*, 98, 175, 1983.
143. **Malaisse, W. J. and Orci, L.,** The role of cytoskeleton in pancreatic B-cell function, *Meth. Achiev. Exp. Pathol.*, 9, 112, 1979.
144. **Rogalski, A. A., Bergmann, J. E., and Singer, S. J.,** Intracellular transport and processing of an integral membrane protein destined for the cell surface is independent of the assembly status of cytoplasmic microtubules, *J. Cell. Biol.*, 95, 337a, 1983.

145. **Rogalski, A. A. and Singer, S. J.,** Association of elements of the Golgi apparatus with microtubules, *J. Cell. Biol.,* 99, 1092, 1984.
146. **Rogalski, A. A., Bergmann, J. E., and Singer, S. J.,** Effect of microtubule assembly status on the intracellular processing and surface expression of an integral protein of the plasma membrane, *J. Cell. Biol.,* 99, 1101, 1984.
147. **Genty, N. and Bussereau, F.,** Is cytoskeleton involved in vesicular stomatitis virus reproduction?, *J. Virol.,* 34, 777, 1980.
148. **Salpeter, M. M. and Farguhar, M. G.,** High resolution analysis of the secretory pathway in mammotrophs of the rat anterior pituitary, *J. Cell. Biol.,* 91, 240, 1981.
149. **Kelly, R. B.,** Pathways of protein secretion in eucaryotes, *Science,* 230, 25, 1985.
150. **Burgess, T. L., Craik, C. S., and Kelly, R. B.,** The exocrine protein trypsinogen is targeted into the secretory granules of an endocrine cell line: studies by gene transfer, *J. Cell. Biol.,* 101, 639, 1985.
151. **Moore, H. P. H., Walker, M. D., Lee, F., and Kelly, R. B.,** Expressing a human proinsulin cDNA in a mouse ACTH-Secreting cell. Intracellular storage, proteolytic processing, and secretion on stimulation, *Cell,* 35, 531, 1983.
152. **Moore, H. P. H. and Kelly, R. B.,** Secretory protein targeting in a pituitary cell line: differential transport of foreign secretory proteins to distinct secretory pathways, *J. Cell. Biol.,* 101, 1773, 1985.
153. **Franc, J. L., Hovsepian, S., Fayet, G., and Bouchilloux, S.,** Differential effects of thyrotropin on various glycosyltransferases in porcine thyroid cells, *Biochem. Biophys. Res. Commun.,* 118, 910, 1984.
154. **Carson, D. D., Earles, B. J., and Lennarz, W. J.,** Enhancement of protein glycosylation in tissue slices by dolichylphosphate, *J. Biol. Chem.,* 256, 11552, 1981.
155. **Sarkar, M. and Mookerjea, S.,** Effect of dexamethasone on mannolipid synthesis by hepatocytes prepared from control and inflamed rats, *Biochem. J.,* 219, 429, 1984.
156. **Kousvelari, E. E., Grant, S. R., and Baum, B. J.,** β-Adrenergic receptor regulation of N-linked protein glycosylation in rat parotid acinar cells, *Proc. Natl. Acad. Sci. U.S.A.,* 80, 7146, 1983.
157. **Kousvelari, E. E., Grant, S. R., Banerjee, D. K., Newby, M. J., and Baum, B. J.,** Cyclic AMP mediates β-adrenergic-induced increases in N-linked protein glycosylation in rat parotid acinar cells, *Biochem. J.,* 222, 17, 1984.
158. **Grant, R. S., Kousvelari, E. E., Banerjee, D. K., and Baum, B. J.,** β-adrenergic stimulation alters oligosaccharide pyrophosphoryldolichol metabolism in rat parotid acinar cells, *Biochem. J.,* 231, 431, 1985.

Chapter 4

TRANSPORT AND STORAGE OF BIOGENIC AMINES IN ADRENAL CHROMAFFIN GRANULES

Shimon Schuldiner

TABLE OF CONTENTS

I. INTRODUCTION

This review will focus on the bioenergetics, mechanism, and molecular basis of biogenic amine transport. During the last few years, direct evidence has been obtained that these processes are coupled chemiosmotically, i.e., the accumulation of neurotransmitters is driven by ion gradients. Two types of neurotransmitter transport systems have been identified: sodium-coupled systems located in the synaptic plasma membrane of nerves, platelets, chromaffin, and glial cells and proton-coupled systems which are part of the membrane of intracellular storage organelles. In several cases besides sodium ions, additional ions, such as chloride and potassium, serve as coupling ions.[1]

The neurotransmitter transport systems seem to be equipped very well to achieve high concentration gradients by utilization of pre-existing ion gradients, created and maintained by ion pumps. Both types of transport systems are able to maintain gradients in the order of 10^4.[1,2] This occurs because it appears that the transporters operate at an effective stoichiometry of several coupling ions per neurotransmitter. In the case of sodium-coupled systems ions participate in a single translocation cycle. In the case of biogenic amine transport into storage organelles, effectively two protons are moving out per amine molecule.

Most of our knowledge about H^+-coupled biogenic amine transport into storage organelles stems from studies of the chromaffin granules.

The isolated granules from the adrenal medulla catalyze uptake of large amounts of adrenaline in an ATP-dependent process[3] which is inhibited by drugs such as reserpine and by proton ionophores.[3,4] The granules have been found to contain a membrane-bound Mg^{2+}-dependent ATPase which is stimulated by proton conductors.[5-7] Evidence has accumulated indicating that the ATPase translocates protons and creates a proton electrochemical gradient across the chromaffin granule membrane. Thus, ATP-dependent proton translocation and ATP-dependent changes in the fluorescence of 1-anilino naftalene-8-sulfonate (ANS)[7] have been detected in intact granules and/or membrane vesicles derived from the granules by osmotic shock. Moreover, it has been demonstrated that an uncoupler-sensitive pH gradient (interior acid) of 2 to 3 pH units is maintained across the chromaffin granule membrane.[8-10] However, the intact granules are not suitable to quantitate the role of pH and/or transmembrane electrical potentials ($\Delta\Psi$) as a driving force for accumulation of biogenic amines. This is mainly due to the fact that the intact granules contain catecholamines at a concentration of 0.8 M together with high concentrations of ATP and bivalent cations. Thus, the amount of free or unbound components in the granules is uncertain. It is therefore difficult to evaluate whether the transport measured in these preparations is indeed against a concentration gradient. This difficulty can be circumvented by using chromaffin granule membrane vesicles derived from the intact granules by osmotic shock.[11,12] These vesicles still catalyze ATP-dependent accumulation of adrenaline and 5-hydroxytryptamine[4,11,12] as well as ATP-dependent proton translocation.[13] Accumulation of adrenaline by chromaffin granule membrane vesicles is dependent upon addition of ATP and inhibited by the appropriate ionophores. Direct evidence has been provided that ΔpH plays a primary role in catecholamine and serotonin transport. It has been shown that a pH gradient generated artificially by a variety of methods can induce reserpine-sensitive biogenic amine transport against its concentration gradient in ghosts (12,14,15) as well as the intact granules.[16] After correlations between $\Delta\Psi$ and uptake of catecholamines had been made in intact granules[17,18] or in ghosts[19-21] it was possible to show that induction of a membrane potential (interior positive), imposed by a potassium gradient ($[K^+]_{out} > [K^+]_{in}$) in the presence of valinomycin, can drive biogenic amine transport or enhance ΔpH driven transport.[22-24] In summary, all available experimental evidence clearly supports the concept that ATP-driven accumulation of biogenic amines is a result of two sequential processes: (1) generation of a proton electrochemical gradient by the membrane-bound ATPase and (2) utilization of $\Delta\mu H^+$ to drive the carrier-mediated accumulation. Since most of the bioenergetic aspects have already been reviewed in several excellent works,[1,38] we will not dwell on these aspects in any detail.

In the coming sections we will discuss some recent kinetic, biochemical, and pharmacological studies. Also, we will present data available on other storage organelles and on the consensus that seems to be evolving about the nature of the H^+-ATPase.

II. THE HYDROGEN ION PUMPING ATPase

It is well established that the acidification of the intravesicular space of chromaffin granules and the establishment of a membrane potential, positive inside, is due to the operation of a H^+-ATPase. Some confusion existed for several years as to the nature of the enzyme. This was mainly due to the fact that most preparations of chromaffin granules contain mitochondrial contaminants. Early findings led researchers to propose that the ATPase of chromaffin granule membranes closely resembled that of mitochondria.[25] However, Cidon and Nelson demonstrated that the presence of this enzyme was most likely due to contamination.[26] Thus, upon treatment of highly purified granule membranes with NaBr all of the F_1-β subunit, as measured with antibodies, dissociates, while up to 80% of the ATPase activity remains associated with the granule.

In an elegant series of experiments Dean et al.[27] characterized the inhibitor profile of the enzyme from chromaffin granules and from platelet dense granules: to avoid complications by other ATPases which might contaminate the preparation, they used an assay that measures only those ATPase molecules which are functionally inserted in the granule membrane and which, therefore, drive accumulation of the biogenic amine. As a control, the ability of an artificially imposed pH gradient to drive transport in the absence of ATP was measured. ATPase inhibitors should not block transport driven by imposed pH gradients, which depend only on the activity of the amine transporter and the relative impermeability of the membrane to H^+. In both preparations the pump activity displays an inhibitor sensitivity distinct from that of mitochondrial F_1F_0 ATPase or Na^+, K^+-ATPase; thus, the enzyme is insensitive to azide, vanadate, oligomycin, ouabain, and efrapeptin. In addition, both NEM (*N*-ethyl-maleimide) and Nbd-Cl (7-chloro-4-nitrobenz-2-oxa-1,3-diazole) preferentially inhibit the granule enzyme. These results suggest that the granule enzyme represents a new class of ATP-driven ion pump very similar to that described in endosomes, lysosomes, coated vesicles, and plant vacuoles.

The chromaffin granule ATPase has been solubilized, reconstituted, and partially purified.[28,29] Four polypeptides seem to be associated with the activity: 115, 72, 57, and 39 kdaltons; the 115- and 57-kdalton peptides are labeled with NEM. Based on the hydrodynamic properties of a detergent-solubilized preparation of chromaffin granules, the mass of the active protein has been estimated to be 134-kdaltons.[30]

A second ATPase associated with granule fractions has been detected.[31] This activity displays a very distinct inhibitor profile: it is insensitive to DCCD or alkylating agents but is strongly inhibited by vanadate. Its structure and function are still obscure.

Many early attempts to purify and reconstitute the granule ATPase have been reported.[32,33] Also, the reversibility of the enzyme has been demonstrated by net ATP synthesis[34] and by ATP-$^{32}P_i$ isotope exhange.[35] An apparent stoichiometry of $2H^+$/ATP has been determined in two laboratories.[36,37] It is not clear as to whether enough care has been taken in all the studies to determine which enzyme is being studied. The pharmacology available at present makes this task possible in the foreseeable future.

For more details with respect to the H^+-ATPase of the chromaffin granule the reader is referred to several recent reviews.[159-161]

III. MECHANISM OF BIOGENIC AMINE TRANSLOCATION

A. Molecular Species of Amine Transported

There now seems to be general agreement that the process of catecholamine uptake is an electrogenic one. This is based on two lines of reasoning: (1) the contribution of the components of $\Delta\mu H^+$ to steady-state amine uptake was measured qualitatively and quantitatively[17-20] and (2) amine uptake was tested upon imposition of artifically generated pH gradient[12-16] and membrane potential.[22-24] In summary, the findings demonstrate that amine uptake is driven by a membrane potential as well as by a pH gradient; the quantitative dependence of the gradient on each of the components is different[20] and fits best to the following equation:

$$\frac{RT}{F} \ln \frac{[\text{amine}]\text{in}}{[\text{amine}]\text{out}} = 2\left(\frac{RT}{F} \ln \frac{[H^+]\text{in}}{[H^+]\text{out}}\right) + \Delta\psi$$

This relation can be accounted for by either of the following: (1) the exchange of an uncharged amine molecule with one proton and a further protonation of the amine in the intravesicular space and (2) exchange of the protonated amine with two protons. Several approaches have been employed to try to distinguish between the two possible mechanisms and thus far there is no clear cut answer. One approach, utilized thus far in several laboratories, has been to test the dependence of the apparent K_m on the medium pH; most of the published findings suggest that there is a steep decrease of the apparent K_m with the increase of the pH.[41,43] When the concentration of the uncharged amine is calculated at each pH, the apparent K_m towards this form does not change, suggesting that the pH dependence observed reflects the change in the concentration of the unchanged species. Similar findings have been reported when binding of dihydrotetrabenazine (TBZOH) (see below) was tested as a function of pH.[44] Further support to the proposal that the uncharged species is the substrate is provided by the finding that a charged analog of epinephrine, dimethylepinephrine, is not a substrate of the transporter.[45] However, in the latter case a methyl group is juxtaposed to the nitrogen of the amine group and this could, by itself, alter the interaction with the protein. Moreover, recently a charged substrate of the system has been reported: *m*-iodobezylguanidine (apparent K_m 2 μM).[46]

B. Asymmetry of the Transporter

Studies performed in the intact organelle indicate a substantial but very slow spontaneous efflux of the endogenous amines even in the absence of exogenous ATP and when the gradients of the amine can be made as large as desired (and favoring efflux) by manipulation of the degree of dilution of the medium.[47-50] Under such conditions (in the absence of ATP) the electrochemical gradient of protons in the intact granule is very low, consisting of a pH gradient (interior acid) and a membrane potential of opposite magnitude (negative inside). Since, on the other hand, the amine gradients are very large it is clear that, under these conditions, the amine is not in thermodynamic equilibrium with the electrochemical gradient of protons.

The slow rate of efflux has been tentatively explained by complexation of the amine by the vesicle core which brings about a decrease in the concentration of the free species.[48] A further analysis of this phenomenon is difficult in the intact granule because of possible binding of the amine to the intravesicular proteins, nucleotides, and ions and also because even the slow leakage of the granule content generates considerable concentrations in the medium.

The reasons for the slow efflux rates have been studied in detail in membrane vesicles depleted from the granule core and from most of the endogenous catecholamine content. 5-

Hydroxytryptamine (5-HT), noradrenaline, and other biogenic amines are accumulated by a preparation of chromaffin granule membrane vesicles against concentration gradients of up to 4000. This uptake is dependent on the presence of ATP and the activity of a H^+-ATPase which generates a proton electrochemical gradient. However, once achieved, the steady-state levels can be maintained for at least 1 hr even in the absence of ATP. Even when the external 5HT is decreased up to 200-fold by dilution, after a steady state is reached, previously accumulated substrate does not efflux to a new steady state as dictated by simple energetic considerations.[57]

These findings would suggest that no input of energy is required to maintain the steady state once this is reached. This would be possible only if there are no considerable leaks under these conditions, i.e., the electrochemical gradient of protons does not decay and the transporter must be catalyzing a 1:1 exchange but no net efflux. In accordance with previous findings, addition of the ionophore nigericin at any time after the steady state has been reached does induce net efflux. It is concluded that the lack of net efflux must be due to a kinetic barrier induced by the acid intravesicular pH: when the intravesicular milieu is acidified below a certain value, the transporter can catalyze exchange of intravesicular amines with extravesicular substrate but it cannot catalyze net efflux of the internal amine. Net efflux, however, can be induced by agents that bring upon an alkalinization of the internal pH, such as nigericin and ammonia salts.

There are several simple models that can explain the translocation cycle of amines in exchange for a hydrogen ion: two of them are schematically depicted in Figure 1. In the top part of the figure a sequential model is presented which is similar to others previously proposed;[16,20] a site on the transporter facing the cytoplasmic side binds substrate with a certain affinity and translocates it to the internal face of the membrane. At this side, the substrate is released and the transporter now becomes protonated. This protonation, which clears the free transporter sites from the intravesicular face of the membrane, is obviously strongly dependent on the pH and is probably a very fast reaction. We assume that (1) the equilibrium of the protonated form between the inside and the outside face and (2) the release of protons in the outside face are not rate-limiting steps since the rate of exchange is high, even when the inside pH is low. Under exchange conditions, the availability of the transporter-binding site inside is made higher because of the continuous flow of solute to the inside. The exchange process will use only the lower part of the upper cycle described in Figure 1 and is almost independent of external pH.

The lower part of Figure 1 describes a simultaneous mechanism in which both the hydrogen ion and the substrate have to be bound to the transporter before translocation occurs. Also in this model binding of protons to the transporter from within will prevent *cis* binding of the amine and therefore efflux. Exchange will occur when external amine binds to the protonated transporter and therefore will not be inhibited by the intravesicular low pH. In fact, if the order of the reaction is as described in the model, binding of substrates to the external side of the membrane should depend on *trans* protonation of the transporter.

Some recent findings seem to support an effect of intravesicular pH on several parameters of the translocation cycle: (1) an increase in the intravesicular pH (at a constant given external pH) brings about an increase in the apparent K_m for amine uptake; (2) reserpine binding is stimulated by an electrochemical gradient of protons;[52,53] and (3) the rate of labeling of the transporter with the specific probe ANPA-5HT (see below) increases with a decrease in the intravesicular pH.[87] Even though these data can be interpreted in a number of ways, they are consistent with both models described in Figure 1.

Regardless of the mechanism, the phenomenon described can, in fact, be considered as a gating device regulated by the intravesicular pH. Part of the mechanism of translocation, the protonation of the transporter in the interior phase of the membrane, is also responsible for the regulation of the process. The pH at which the gate will shut off depends on the pK

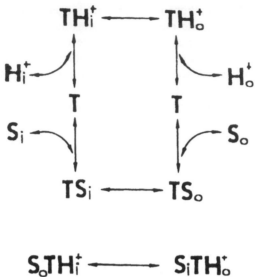

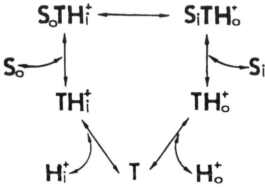

FIGURE 1.   Schematic description of a translocation cycle.
Top: a sequential model; bottom: a simultaneous model. T,
transporter; S, substrate. (From Schuldiner, S. et al., *Ann.
N.Y. Acad. Sci.*, 456, 268, 1985. With permission.)

of one or more residues on the protein. This type of gating mechanism has since been described in at least two other transport systems: the Na^+/H^+ antiporter from eukaryotes[54] and prokaryotes.[55]

The possibility that the unprotonated species of 5HT is highly permeant through the membrane has been considered. Indeed, there have been reports that liposomes prepared with purified soybean lipids display a high permeability to the unprotonated form of various biogenic amines.[56,57] However, an interesting corollary of the findings is that this does not seem to be the case with the lipids from the chromaffin granules since dilution of the extravesicular medium does not induce efflux even when the intravesicular concentrations are up to 5 mM 5HT.[57] Exchange, on the other hand, is quite rapid under the conditions where there is no net efflux. Also, photoinactivation of the transporter with ANPA-5HT significantly slows down the rate of nigericin-induced efflux.[51] Furthermore, if the protonated amine were permeable through the membrane, it would be in equilibrium with the pH gradient. It has already been shown that this is not the case.[2] The low permeability of 5HT is also shared by other membranes such as the cytoplasmic one in platelets, in which downhill movement of the amine requires Na^+ and Cl^-, the co-transported ions in this system, and is inhibited by imipramine, an inhibitor of the sodium-coupled serotonin transporter.[58]

The finding that there is a rapid exchange under conditions which are apparently a steady state are in agreement with the possibility that the various transmitters, dopamine, norepinephrine, and epinephrine, are synthesized on opposite sides of the membrane.[2] Recently, it has even been suggested that ATP may have a regulatory role on the exchange process.[50]

The significance of the findings described for the secretory cell is self-evident: the energetic burden of maintaining concentrations of amines of up to 0.6 to 0.7 M at μM cytoplasmic levels[2] could be prohibitive. Moreover, leakage of biologically active amines into the cytoplasm could have serious effects on the metabolism of the cell. In a steady-state system with no "gating" device, this would occur every time the levels of ATP in the cytoplasm decrease. A loss of intravesicular material would occur also upon decrease of the cytoplasmic levels of the amine.

Regardless of the mechanism, one of the most straightforward predictions of these findings is that in vivo inhibitors of the synthesis of the amines should not cause depletion of the stores. Indeed, even inhibition of synthesis by α-methyltyrosine, at the normally rate-limiting step, that of tyrosine hydroxylation, does not alter the catecholamine content of the adrenal medulla at rest.[59] The half-life of adrenaline and noradrenaline in different tissues ranges from 4 hr for brain to 300 hr for the adrenal medulla.[60] It has been shown that reserpine is a competitive and reversible inhibitor of amine influx.[23] Early experiments indicate enhancement of efflux by reserpine. However, these experiments are complicated by a nonspecific detergent-like effect at the ratio of reserpine-protein used.[61] Appreciation of this difficulty, together with our better understanding of the efflux process obtained in this study, should be helpful in the clarification of the mechanism of action of reserpine.

Reserpic acid, an impermeant derivative of reserpine, inhibits uptake from the outside but not from the inside.[62] This finding has been interpreted as evidence that the transporter is also structurally asymmetrical.

IV. MOLECULAR PHARMACOLOGY OF THE TRANSPORTER

Significant progress towards an understanding of the pharmacology of the amine transporter at a molecular level has been made during the past 5 years. Thus, several probes have been synthesized and their binding has been studied to learn about the properties of the binding site(s) and to identify the putative subunits of the transporter.

A. Tetrabenazine Binding

[3H]-Dihydrotetrabenazine (TBZOH), a derivative of the drug tetrabenazine, was synthesized.[63] TBZOH binds to a single site on the chromaffin granule membrane with an apparent binding constant of about 3 nM. The density of the TBZOH binding site is about 60 pmol/mg protein. The K_d is very similar to the apparent K_i of the amine transport. Also, binding was correlated with inhibition of norepinephrine transport. However, displacement of TBZOH from its binding site by transport substrates such as 5HT or NE is achieved only at concentrations about 100-fold higher than the apparent K_m of the respective compounds.[53,63] Reserpine does not inhibit binding at concentrations which completely inhibit transport. At higher concentrations (K_i = 20 nM) however, it does inhibit binding as well.

The component responsible for the binding has been solubilized by treatment with detergents: sodium cholate solubilized more than 70% of the binding sites.[64] High detergent concentrations had a reversible inhibitory effect. The binding characteristics (K_d = 23 nM, B_{max} = 90 pmol/mg protein) and the pharmacological properties of the binding sites were also similar to those of the membranes.

Radiation inactivation experiments have shown that the apparent M_r of the TBZ binding protein is 65 kdaltons.[65]

B. Reserpine Binding

[3H]-Reserpine was used to measure binding of this drug to granule membranes. Weaver and Deupree[52] and Scherman and Henry[53] found that reserpine binds to membranes energized by the addition of ATP and is inhibited by protonophores. Scherman and Henry showed that binding also occurs in the absence of ATP and ATP accelerates the rate of binding without effecting plateau values and equilibrium constants. They found that binding occurs on two classes of sites: R_1, B_{max} = 7 pmol/mg protein and K_d = 0.7 nM and R_2, B_{max} = 60 pmol/mg protein and K_d = 25 nM. Deupree and Weaver[66] found only one class of sites with intermediate affinity and low density (B_{max} = 7.8 pmol/mg protein and K_d = 9 nM). Scherman and Henry[53] propose that sites R_2 are equivalent to TBZOH binding sites as the densities are similar and because TBZ displaces reserpine from R_2 sites. Sites R_1, on the other hand, are TBZ resistant and they seem to be involved in amine uptake since K_d values are similar to the K_i values of reserpine for NE uptake. Moreover, transport substrates displace reserpine from R_1 sites at concentrations similar to their apparent K_m values.

Radiation inactivation experiments show that the apparent M_r of the reserpine binding component is 40 kdaltons.[65]

C. Photoaffinity Labels: Azido Derivatives of 5HT

Three photoactive derivatives have been used to identify the transporter of bovine chromaffin granules:[67] pig platelet storage organelles, rat brain synaptic vesicles, and 5-HT-containing organelles in rat basophilic leukemia cells.[88]

The specificity of the first one, 4-azido,3-nitrophenylazo-[5HT] (ANPA-5HT), is supported by the following criteria. (1) ANPA-5HT is a competitive inhibitor of amine transport in the dark. The apparent K_i is identical to the apparent K_m of transport of 5-HT, its parent compound.[67] The latter indicates that the transporter recognizes the modified substrate. (2) Upon illumination, ANPA-5HT photoinactivates the amine transport. Transport of unrelated neurotransmitters or the generation of a pH gradient across the membrane is not inhibited.[67] (3) The rate of photoinactivation is lower in the presence of other substrates of the transporter. The concentrations required to ''protect'' the transporter correlate well with the known affinities of the various substrates. From the dependence of the rate of photoinactivation on the concentration of ANPA-5HT an apparent kinetic constant can be calculated. This constant is similar to the apparent K_i. The agreement suggests that the inactivation of transport is a result of a sequence of events which starts when ANPA-5HT binds to the transporter and is thereafter activated by light. (4) The rate of labeling of a membrane suspension with [³H]-ANPA-5HT in the light is lower in the presence of other substrates of the transporter.

Recently, ANPA-5HT has been iodinated with ¹²⁵I.[89] The resulting compound displays properties almost identical to those of ANPA-5HT and dramatically shortens the time required for visualization of the label.

The pattern of labeling of the membrane polypeptides has been analyzed by separating them by SDS-gel electrophoresis and subjecting the gel to fluorography or autoradiography. More than 80% of the label in the protein fraction is associated with a 48-kdalton polypeptide and the labeling is inhibited by reserpine and by other substrates of the transporter. The concentration of reserpine required to inhibit labeling of 50% is similar to its apparent K_i transport.

The characteristics of the specific labeling of the amine transporter have been studied by assessment of the amount of radioactivity associated with the membrane proteins under several conditions. The criterion of the specificity of the labeling is that when reserpine or any of the other four transport substrates tested thus far, 5-hydroxytryptamine, epinephrine, norepinephrine, and dopamine are present during illumination at concentrations above their K_m, the amount of labeling is dramatically decreased.[67,68]

Since transmembrane gradients are not needed for labeling of the transporter with ANPA-5HT,[68] we tested whether labeling requires at all the presence of an intact membrane. Extraction of chromaffin granule membranes with 1% cholate renders a soluble fraction that can be reconstituted into liposomes.[57] The proteoliposomes thus obtained display the transporter activity. We have now found that even in the presence of the detergent, a protein is still specifically labeled by ANPA-5HT: however, labeling of the 48-kdalton polypeptide is inhibited at concentrations above 0.5%. A second polypeptide (apparent M_r 56 kdaltons) is labeled under these conditions. The concentrations of the substrates required to prevent labeling are higher than those required in the intact membrane. It is worth mentioning that in at least two other detergents, octylglucoside and Triton X-100, labeling of the 48-kdalton peptide is not inhibited even at concentrations that achieve full solubilization.[90]

D. Azido Derivative of Tetrabenazine

An azido derivative of tetrabenazine, a specific inhibitor of transport, has been synthesized.[69] The compound, [³H]TBA, binds reversibly to the granule membranes in the dark with a K_d of about 50 nM and a density of sites of 40 to 50 pmol/mg protein, consistent with reported densities of reserpine and dihydrotetrabenazine binding sites. Upon irradiation, TBA bound irreversibly to a polypeptide with an apparent M_r of 70,000. Since TBA and TBZ compete for the same binding site (R2 type), the authors suggest that the 70-kdalton polypeptide is an R2 binding site.[69] This suggestion is supported by radiation inactivation studies in which it was found that the apparent M_r of the TBZ binding site is 65 kdaltons.[65]

V. CHEMICAL MODIFICATIONS

Despite the profusion of noncovalent inhibitors and ligands of the catecholamine transporter, very little information has been gathered about covalent modifiers of the protein. The native protein does not seem to have essential sulfhydryl groups accessible to NEM or pCMB. DCCD inhibits transport[12,70] and this inhibition can be prevented by dihydrotetrabenazine.[70] Diethylpyrocarbonate, which under proper conditions reacts specifically with histidine moieties, inhibits the transporter by lowering its V_m, without alteration of the K_m.[71] Interestingly, binding of dihydrotetrabenazine is not hampered.

VI. RECONSTITUTION OF THE AMINE TRANSPORTER

The amine transporter from bovine chromaffin granules has been solubilized with sodium cholate in the presence of soybean phospholipids.[57,72] The solubilized protein has been incorporated into liposomes after removal of the detergent either by gel filtration or by dialysis. Reserpine- and tetrabenazine-sensitive accumulation against concentration gradients has been achieved by imposition of artificially generated pH gradients and membrane potential. The apparent K_m of the reconstituted transporter is similar or even lower than that of the native one. Two major problems have been encountered in the reconstitution of transport activity: (1) it is necessary to add asolectin prior to the detergent to maintain a fully functional transporter. Interestingly, this requirement is superseded when crude bovine brain lipids are added at reconstitution; (2) a large reserpine-insensitive accumulation is detected either in bare liposomes or in reconstituted proteoliposomes.[57] This is due to the relatively high passive permeability of the various amines across most membranes and subsequent equilibration with the imposed pH gradients. A further complication arises from the fact that reserpine is a very hydrophobic compound with detergent-like properties. Thus, at high concentrations it will induce leakiness even in liposomes and inhibit even unmediated transport.[61] Therefore, very rigorous criteria should be applied in every single reconstitution experiment to distinguish between mediated and unmediated transport.

VII. PURIFICATION OF THE TRANSPORTER

The putative amine transporter from bovine adrenal chromaffin granules has been partially purified in a single step utilizing affinity chromatography.[73] A 5-hydroxytryptamine moiety has been coupled to a Sepharose 4B matrix in a position ortho to the hydroxyl group. When membranes solubilized with sodium cholate are chromatographed on the above matrix a polypeptide is highly enriched. The enrichment is dependent on the presence of the proper ligand on the matrix and is inhibited if the column is previously equilibrated with a soluble ligand. Enrichment of the above polypeptide is accompanied by an increase in the specific activity of the transporter as measured by its labeling by 4-azido,3-nitrophenyl-azo-(5-hydroxytryptamine). The ability of reserpine, a competitive inhibitor of binding and transport, to inhibit labeling of the purified transporter correlates well with its known kinetic constants in the native membranes.

The polypeptide purified was first thought to be identical to the one previously identified as the putative transporter based on specific labeling by a photoaffinity label.[73] However, a more detailed analysis of the electrophoretic pattern and the realization that: (1) in the presence of 1% sodium cholate, a 56-kdalton peptide but not the 48-kdalton peptide is labeled with ^{125}I-ANPA-5HT and (2) an antibody raised against the purified polypeptide cross-reacts, both in the membrane and in cholate extracts with a protein that has a mobility of the 56-kdalton and not 48-kdalton polypeptide.[74]

VIII. AMINE STORAGE IN OTHER ORGANELLES

In addition to the catecholamine-rich chromaffin granules of adrenal medulla, other intracellular organelles from a wide variety of secretory cells accumulate biogenic amines. These include the adrenergic synaptic vesicles,[75-77] 5-hydroxytryptamine containing secretory granules in platelets, enterochromaffin cells, and serotonergic neurons;[78-80] dopamine-rich vesicles in dopaminergic neurons;[81] and histamine-containing granules in mast cells and basophils.[82] Where it has been studied, the mechanism of amine accumulation into isolated storage organelles or membrane vesicles derived therefrom always involves exchange of extravesicular amine with intravesicular H^+, catalyzed by a reserpine-sensitive transporter.[76,77,83,84] An H^+-pumping ATPase in the organelle membrane generates an electrochemical gradient of H^+ (acid and positive inside) which provides the driving force for amine-H^+ exchange. All the organelles studied display the same inhibitor sensitivity, irrespective of the amine normally stored within.[85] These findings have led to the proposal that a closely similar or identical protein catalyzes amine-H^+ exchange in all biogenic amine storage organelles.[83]

Binding of TBZOH and the apparent K_m for 5HT transport were measured in mice brain synaptic vesicles.[86] The values obtained (3 nM and 0.8 μM) are very similar to the values measured in adrenal chromaffin granules.

[^{3}H]-Reserpine binding was measured in a synaptic vesicle preparation from bovine caudate nucleus.[83] Binding is of a high affinity type (K_d = 1.2 nM, B_{max} = 3.3 pmol/mg protein) and is dependent on ATP and inhibited by protonophores. Substrates displace reserpine at concentrations similar to those required for inhibition of dopamine transport.

ANPA-5HT was used in an attempt to identify the transporter in various organelles:[74] it inhibits ATP-driven reserpine-sensitive 5HT transport into membrane vesicles prepared from porcine platelet dense (PL) granules, rat brain synaptic vesicles (SV), and histamine-containing granules of rat basophilic leukemia (RBL) cells. In addition, it specifically labels a polypeptide in each of the above-mentioned preparations. The apparent molecular weight of the labeled protein band varies from approximately 52 to 34-kdaltons depending on the source of vesicles. The largest variation occurs between SV and RBL vesicles, while in both PL and chromaffin granule (CG) vesicles a 48-kdalton polypeptide is labeled. This result indicates that the molecular weight variation is not due to species difference, since RBL and SV membranes are both derived from rat, while CG and PL vesicles are from cow and pig, respectively. A more likely explanation is that the differences result from either proteolysis or processing or both, since the 34-kdalton band is occasionally observed in overexposed samples of labeled CG vesicles. The 52-kdalton band may be similar to the 56-kdalton band labeled in the presence of cholate in chromaffin granules. Still, the possibility that the differences reflect functionally similar but structurally different amine transporters cannot be ruled out at present. Analysis of the labeled polypeptides by peptide mapping is likely to help in resolving this issue.

ACKNOWLEDGMENT

The studies performed in our lab were supported by NIH grant, NS 16708. I wish to thank all my collleagues who sent reprints and preprints prior to publication.

REFERENCES

1. **Kanner, B. I.,** Bioenergetics of neurotransmitter transport, *Biochim. Biophys. Acta,* 726, 293, 1983.
2. **Njus, D., Knoth, J., and Zallakian, M.,** Proton-linked transport in chromaffin granules, *Curr. Top. Bioenerg.,* 11, 107, 1981.
3. **Kirshner, N.,** Uptake of catecholamine by a particular fraction of the adrenal medulla, *J. Biol. Chem.,* 237, 2311, 1962.
4. **Bashford, C. L., Casey, R. P., Radda, G. K., and Ratchie, G. A.,** The effect of uncouplers on catecholamine incorporation by vesicles of chromaffin granules, *Biochem. J.,* 148, 153, 1975.
5. **Banks, R.,** The adenosine-triphosphatase activity of adrenal chromaffin granules, *Biochem. J.,* 95, 490, 1965.
6. **Hasselbach, W. and Taugner, G.,** The effect of a cross-bridging thiol reagent on the catecholamine fluxes of adrenal medulla vesicles, *Bochem. J.,* 119, 265, 1970.
7. **Bashford, C. L., Radda, G. K., and Ritchie, G. A.,** Energy-linked activities of the chromaffin granule membrane, *FEBS Lett.,* 50, 21, 1975.
8. **Casey, R. P., Njus, D., Radda, G. K., and Sehr, P. A.,** Active proton uptake by chromaffin granules: observation by amine distribution and phosphorus-31-nuclear magnetic resonance techniques, *Biochemistry,* 16, 972, 1977.
9. **Johnson, R. G. and Scarpa, A.,** Internal pH of isolated chromaffin vesicles, *J. Biol. Chem.,* 251, 2189, 1976.
10. **Pollard, H. B., Zinder, O., Hoffman, P. G., and Nikodejenic, O.,** Regulation of the transmembrane potential of isolated chromaffin granules by ATP, ATP analogs and external pH, *J. Biol. Chem.,* 251, 4544, 1976.
11. **Phillips, J. H.,** Transport of catecholamines by resealed chromaffin-granule ghosts, *Biochem. J.,* 144, 311, 1975.
12. **Schuldiner, S., Fishkes, H., and Kanner, B. I.,** Role of a transmembrane pH gradient in epinephrine transport by chromaffin granule membrane vesicles, *Proc. Natl. Acad. Sci. U.S.A.,* 75, 3713, 1978.
13. **Flatmark, T. and Ingebresten, O. C.,** ATP-dependent proton translocation in resealed chromaffin granule ghosts, *FEBS Lett.,* 78, 53, 1977.
14. **Ingebretsen, O. C. and Flatmark, T.,** Active and passive transport of dopamine in chromaffin granule ghosts isolated from bovine adrenal medulla, *J. Biol. Chem.,* 254, 3833, 1979.
15. **Phillips, J. H.,** 5 Hydroxytryptamine transport by the bovine chromaffin-granule membrane, *Biochem. J.,* 70, 673, 1978.
16. **Johnson, R. G., Carlson, N., and Scarpa, A.,** ΔpH and catecholamine distribution in isolated chromaffin granules, *J. Biol. Chem.,* 253, 1512, 1978.
17. **Holz, R. W.,** Evidence that catecholamine transport into chromaffin vesicles is coupled to vesicle membrane potential, *Proc. Natl. Acad. Sci. U.S.A.,* 75, 5190, 1978.
18. **Johnson, R. G. and Scarpa, A.,** Proton motive force and catecholamine transport in isolated chromaffion granules, *J. Biol. Chem.,* 254, 3750, 1979.
19. **Johnson, R. G., Pfister, D., Carty, S. E. and Scarpa, A.,** Biological amine transport in chromaffin ghosts-coupling to the transmembrane proton and potential gradients, *J. Biol. Chem.,* 254, 10963, 1979.

20. **Knoth, L. J., Handloser, K., and Njus, D.,** Electrogenic epinephrine transport in chromaffin granules ghosts, *Biochemistry,* 19, 2938, 1980.

21. **Drake, R. A. L., Harvey, S. A. K., Njus, D., and Radda, G. K.,** The effect of chlorpromazine on bioenergetic processes in chromaffin granule membranes, *Neuroscience,* 4, 853, 1979.

22. **Njus, D. and Radda, G. K.,** A potassium ion diffusion potential causes adrenaline uptake in chromaffin-granule ghosts, *Biochem. J.,* 180, 575, 1979.

23. **Kanner, B. I., Sharon, I., Maron, R., and Schuldiner, S.,** Electrogenic transport of biogenic amines in chromaffin granule membrane vesicles, *FEBS Lett.,* 111, 83, 1980.

24. **Apps, D. K., Pride, J. G., and Phillips, J. H.,** Both the transmembrane pH gradient and the membrane potential are important in the accumulation of amines by resealed chromaffin-granule "ghosts", *FEBS Lett.,* 111, 386, 1980.

25. **Apps, D. and Schatz, G.,** An adenosine triphosphatase isolated from chromaffin-granule membranes is closely similar to F_1-adenosine triphosphatase of mitochondria, *Eur. J. Biochem.,* 100, 411, 1979.

26. **Cidon, S. and Nelson, N.,** A novel ATPase in the chromaffin granule membrane, *J. Biol. Chem.,* 258, 2892, 1983.

27. **Dean, G. E., Fishkes, H., Nelson, P. J., and Rudnick, G.,** The hydrogen ion pumping ATPase of platelet dense granule membranes, *J. Biol. Chem.,* 259, 9569, 1984.

28. **Cidon, S., Ben David, H., and Nelson, N.,** ATP-driven proton fluxes across membranes of secretory organelles, *J. Biol. Chem.,* 258, 11684, 1983.

29. **Nelson, N. and Cidon, S.,** Chromaffin granule proton pump, *Methods Enzymol.,* in press.

30. **Rudnick, G.,** Hydrodynamic properties of the chromaffin granule: hydrogen ion pumping ATPase, *Biochemistry,* in press.

31. **Percy, J. M., Pryde, J. G., and Apps, D. K.,** Isolation of ATPase I, the proton pump of chromaffin granule membranes, *Biochem. J.,* 231, 1985.

32. **Buckland, R. M., Radda, G. K., and Wakefield, L. M.,** Reconstitution of the Mg^{2+} ATPase of the chromaffin granule membrane, *FEBS Lett.,* 323, 1979.

33. **Giraudaut, J., Roisin, M. P., and Henry, J. P.,** Solubilization and reconstitution of the adenosine 5'-triphosphate dependent proton translocase of bovine chromaffin granule membrane, *Biochemistry,* 19, 4499, 1980.

34. **Roisin, M. P., Scherman, D., and Henry, J. P.,** Synthesis of ATP by an artificially imposed electro-chemical gradient in chromaffin granule ghosts, *FEBS Lett.,* 115, 143, 1980.

35. **Roisin, M. P. and Henry, J. P.,** Purification and reconstitution of the [323]P-ATP exchange activity of bovine chromaffin granule membrane, *Biochim. Biophys. Acta,* 681, 282, 1982.

36. **Njus, D., Sehr, P. A., Radda, G. K., Ritchie, G. M., and Seeley, P. J.,** Phosphorus-31 nuclear magnetic resonance studies of active proton translocation in chromaffin granules, *Biochemistry,* 17, 4337, 1978.

37. **Johnson, R. G., Beers, M. F., and Scarpa, S.,** H⁺-ATPase of chromaffin granules. Kinetics, regulation and stoichiometry, *J. Biol. Chem.,* 257, 10701, 1982.

38. **Winkler, H., Apps, D. K., and Fischer-Colbrie, R.,** The molecular function of adrenal chromaffin granules: established facts and controversial results, *Neuroscience,* in press.

39. **Rudnick, G.,** ATP-driven H⁺ pumping into intracellular organelles, *Ann. Rev. Physiol.,* 48, 403, 1986.

40. **Rudnick, G.,** Acidification of intracellular organelles: mechanism and function, *Physiol. Membrane Disorders,* 25, 1986.

41. **Scherman, D. and Henry, J. P.,** pH Dependence of the ATP-driven uptake of noradrenaline by bovine chromaffin granule ghosts, *Eur. J. Biochem.,* 116, 535, 1981.

42. **Johnson, R. G., Carty, S. E., and Scarpa, A.,** Coupling of H⁺ gradients to catecholamine granules, *Ann. N.Y. Acad. Sci.,* 456, 254, 1985.

43. **Knoth, J., Isaacs, J. M., and Njus, D.,** Amine transporter chromaffin granule ghosts pH dependence implies cationic form is translocated, *J. Biol. Chem.,* 256, 6541, 1981.

44. **Scherman, D. and Henry, J. P.,** The catecholamine carrier of bovine chromaffin granules form of the bound amine, *Mol. Pharmacol.,* 23, 431, 1983.

45. **Ramu, A., Levine, M., and Pollard, H.,** Chemical evidence that catecholamines are transported across the chromaffin granule membrane as noncationic species, *Proc. Natl. Acad. Sci. U.S.A.,* 80, 2107, 1983.

46. **Desplanches, G. and Henry, J. P.,** Uptake of metaiodobenzylguanidine by bovine chromaffin granule membranes, *Mol. Pharmacol.,* 29, 275, 1986.

47. **Blaschko, H., Hagen, P., and Welch, A. D.,** Observations on the intracellular granules of the adrenal medulla, *J. Physiol. (London),* 129, 27, 1955.

48. **Hillarp, N. A.,** Different pools of catecholamines stored in the adrenal medulla, *Acta Physiol. Scand.,* 50, 8, 1960.

49. **Lishajko, F.,** Release, reuptake and net uptake of dopamine, noradrenaline and adrenaline in isolated sheep adrenal medulary granules, *Acta Physiol. Scand.,* 76, 159, 1969.

50. **Ramu, A. and Pollard, H. B.,** ATP-activated exchange of catecholamines by isolated intact chromaffin granules, Minisymposium Transport Mechanisms in Organelles, *Fed. Proc.,* 41, 2755, 1982.

51. **Maron, R., Stern, Y., Kanner, B. I., and Schuldiner, S.,** Functional asymmetry of the amine transport from chromaffin granules, *J. Biol. Chem.,* 258, 11476, 1983.

52. **Weaver, J. H. and Deupree, J. D.,** Conditions required for reserpine binding to the catecholamine transporter on chromaffin granule ghosts, *Eur. J. Pharmacol.,* 80, 437, 1982.

53. **Scherman, D. and Henry, J. P.,** Reserpine binding to bovine chromaffin granule membranes — characterization and comparison with dihydrotetrabenazine binding, *Mol. Pharmacol.,* 25, 113, 1984.

54. **Aronson, P.,** Kinetic properties of the Na$^+$/H$^+$ exchanger, *Annu. Rev. Physiol.,* 47, 545, 1985.

55. **Padan, E. and Schuldiner, S.,** Intracellular pH and membrane potentials as regulators in the prokaryotic cell, *J. Membrane Biol.,* in press.

56. **Nichols, J. and Deamer, D.,** Catecholamine accumulation by liposomes maintaining pH gradients, *Biophys. J.,* 17, 183a, 1977.

57. **Maron, R., Fishkes, H., Kanner, B. I., and Schuldiner, S.,** Solubilization and reconstitution of the catecholamine transporter from chromaffin granules, *Biochemistry,* 18, 4781, 1979.

58. **Nelson, P. J. and Rudnick, G.,** Coupling between platelet 5-hydroxytryptamine and potassium transport, *J. Biol. Chem.,* 254, 10084, 1979.

59. **Gordon, R., Spector, S., Sjoerdsma, A., and Udenfriend, S.,** Increased synthesis of norepinephrine and epinephrine in the intact rat during exercise and exposure to cold, *J. Pharmacol. Exp. Ther.,* 153, 440, 1966.

60. **Udenfriend, S.,** Biosynthesis of the sympathetic neurotransmitter, norepinephrine, *Harvey Lect.,* 60, 57, 1964.

61. **Zallakian, M., Knoth, J., Metropoulous, G. E., and Njus, D.,** Multiple effects of reserpine on chromaffin-granule membranes, *Biochemistry,* 21, 1051, 1982.

62. **Chaplin, L., Cohen, A. H., Heuttl, P., Kennedy, M., Njus, D.,** Multiple effects of reserpine on chromaffin-granule membranes, *Biochemistry,* 21, 1051, 1982.

63. **Scherman, D., Jauodon, P., and Henry, J. P.,** Characterization of the monoamine carrier of chromaffin granule membrane by binding of ^{3}H dihydrotetrabenazine, *Proc. Natl. Acad. Sci. U.S.A.,* 80, 584, 1983.

64. **Scherman, D. and Henry, J. P.,** Solubilization of the catecholamine carrier of chromaffin granule membranes in a form that binds substrates and inhibitors of uptake, *Biochemistry,* 22, 2805, 1983.

65. **Henry, J. P.,** personal communication.

66. **Dupree, J. D. and Weaver, J. A.,** Identification and characterization of the catecholamine transporter in bovine chromaffin granules using ^{3}H-reserpine, *J. Biol. Chem.,* 259, 10907, 1984.

67. **Gabizon, R., Yetinson, T., and Schuldiner, S.,** Photoinactivation and identification of the biogenic amine transporter in chromaffin granules from bovine adrenal medulla, *J. Biol. Chem.,* 257, 15145, 1982.

68. **Schuldiner, S., Gabizon, R., Maron, R., Suchi, R., and Stern, Y.,** The amine transporter from bovine chromaffin granules, *Ann. N.Y. Acad. Sci.,* 456, 268, 1985.

69. **Isambert, M. F. and Henry, J. P.,** Photoaffinity labeling of the tetrabenazine binding sites of bovine chromaffin granule membranes, *Biochemistry,* 24, 3660, 1985.

70. **Gasnier, B., Scherman, D., and Henry, J. P.,** Dicyclohexycarbodiimide inhibits the monoamine carrier of bovine chromaffin granule membrane, *Biochemistry,* 24, 1239, 1985.

71. **Isambert, M. F. and Henry, J. P.,** Effect of diethylpyrocarbonate on pH driven monoamine uptake by chromaffin granule ghosts, *FEBS Lett.,* 136, 13, 1981.

72. **Isambert, M. F. and Henry, J. P.,** Solubilization and reconstitution of ATP-dependent noradrenaline uptake system of bovine chromaffin granule membrane, *Biochimie,* 63, 211, 1981.

73. **Gabizon, R. and Schuldiner, S.,** The amine transporter from bovine chromaffin granules, *J. Biol. Chem.,* 260, 3001, 1985.

74. **Schuldiner, S., Gabizon, R., Stern, Y., and Suchi, R.,** The amine transporter from bovine chromaffin granules: photolabeling and partial purification, *Ann. N.Y. Acad. Sci.,* 493, 189, 1987.

75. **Kirshner, N.,** Molecular organization of the chromaffin vesicles of the adrenal medulla, *Adv. Cytopharmacol.,* 2, 265, 1974.

76. **Toll, L. and Howard, B. D.,** Role of Mg^{2+}-ATPase and a pH gradient in the storage of catecholamines in synaptic vesicles, *Biochemistry,* 17, 2517, 1978.

77. **Maron, R., Kanner, B. I., and Schuldiner, S.,** The role of a transmembrane pH gradient in 5-hydroxytryptamine uptake by synaptic vesicles from rat brain, *FEBS Lett.,* 98, 237, 1979.

78. **Tranzer, J. P., DaPrada, M., and Pletscher, A.,** Ultrastructural localization of 5-hydroxytryptamine in blood platelets, *Nature (London),* 212, 1574, 1966.

79. **Da Prada, M. and Pletscher, A.,** Differential uptake of biogenic amines by isolated 5-hydroxytryptamine organelles of blood platelets, *Life Sci.,* 8, 65, 1969.

80. **Beaven, M. A., Soll, A. H., and Lewin, K. J.,** Histamine synthesis by intact masts cells from canine fundic mucosa and liver, *Gastroenterology,* 82, 254, 1982.

81. **Whittaker, V. P.,** Subcellular localization of neurotransmitters, *Adv. Cytopharmacol.,* 1, 319, 1971.

82. **Mota, I., Beraldo, W. T., Ferri, A. G. and Junqueira, L. V.,** Intracellular distribution of histamine, *Nature (London),* 174, 698, 1954.

83. **Rudnick, G., Fishkes, H., Nelson, P. J., and Schuldiner, S.,** Evidence for two distinct serotonin transport systems in platelets, *J. Biol. Chem.,* 255, 3638, 1980.
84. **Fishkes, H. and Rudnick, G.,** Bioenergetics of serotonin transport by membrane vesicles derived from platelet dense granules, *J. Biol. Chem.,* 257, 5671, 1982.
85. **Iverson, L. L.,** Uptake processes for biogenic amines, in *Handbook of Psychopharmacology,* Vol. 3, Iversen, L. L., Iversen, S. D., and Snyder, S. H., Eds., Plenum Press, New York, 1975, 381.
86. **Scherman, D. and Henry, J. P.,** unpublished results.
87. **Stern, Y. and Schuldiner, S.,** unpublished observations.
88. **Gabizon, R. et al.,** in preparation.
89. **Suchi, R. and Schuldiner, S.,** in preparation.
90. **Schuldiner, S. et al.,** unpublished observations.

Chapter 5

MAINTENANCE OF RESPONSIVITY

Adrie J. M. Verhoeven

TABLE OF CONTENTS

I. INTRODUCTION

When properly stimulated, secretory cells undergo a number of changes which eventually lead to the extrusion of granule-stored substances. The cell's responsivity to a secretagogue is determined by several factors. In order to become stimulated the cell must maintain its whole spectrum of receptors available and the mechanisms that couple receptor-occupancy to the extrusion of granule contents in a sensitive but *dormant* state. Once an agonist is bound to its specific receptor, this initial signal must be transduced into (a set of) second messengers that initiate the machinery responsible for the relative movement of the granules towards the plasma membrane. Finally, the granule and plasma membranes must fuse in order to promote externalization of the stored material. The responsivity of a cell is thus determined by the amount of agonist that is required to initiate a response (sensitivity) and the maximal response that can be obtained (response potential).

Secretion is, by definition,[1] an energy-requiring process. This is illustrated by its susceptibility to metabolic inhibitors and the concurrent acceleration of ATP-regenerating sequences. The involvement of energy in secretory response is also indicated by studies with permeabilized cells; these cells require a hydrolyzable nucleoside-triphosphate to function normally.[2,3] Energy may be required in the actual secretion process itself, at several stages in the signal-processing sequence, as well as at the prestimulatory stage for the maintenance of the cell's responsivity. Whereas the remainder of this book addresses the energetics of the stimulated secretory cell, this chapter will focus on the unstimulated cell. Blood platelets are central in this chapter, since of all secretory cells they are the most extensively studied in this respect. Platelets are especially suited for this kind of study, since they have a relatively simple energy metabolism, and can be stimulated by a variety of agonists to perform a number of well-defined secretory responses.

II. ENERGY STATUS AND RESPONSIVITY

Several ATP-consuming steps have now been identified in the sequence of stimulus-secretion coupling in various secretory cells, such as the phosphorylation of phosphatidyl-inositol (PI) and diacylglycerol (DAG), adenylate cyclase, protein kinases, and actomyosin ATPase. Both in platelets[4-6] and pancreatic β-cells,[7] secretion responses and changes in ATP-turnover follow similar kinetics; at least in platelets, the progression of secretion is interrupted as soon as cytoplasmic ATP is exhausted.[8] Hence, simultaneous ATP-consumption is essential for the induction and execution phase of secretion. At the moment of stimulation, the cell must therefore, have a high level of cytoplasmic ATP and/or a high ATP-regenerating capacity to cope with the energy demands of secretion. An additional requirement was indicated in a study on platelet acid hydrolase secretion:[4] platelets that were ATP-depleted to various extents before stimulation showed much higher energy costs for this response than platelets whose ATP levels were preserved until the moment of stimulation. This extra energy requirement may be related to maintaining the cells in an optimally responsive state.

In spite of the energy dependency, a 50% fall in cytoplasmic ATP can be induced in unstimulated platelets without affecting responsivity (Reference 9 and citations therein). Under these conditions, however, the adenylate energy charge

$$AEC = (ATP + 0.5\ ADP)/(ATP + ADP + AMP)$$

is preserved at values higher than 0.88. A high AEC is thought to reflect a high cellular energy state.[10] These observations therefore suggest that AEC, rather than the ATP level, governs energy availability for the secretory responses.

The concept of the AEC as an overall energy state indicator is still not generally accepted.[11] A more comprehensible parameter is the phosphorylation potential ($(P_i \cdot (ADP)/ATP)$) which directly reflects the amount of free energy that is stored in the cytoplasmic adenine nucleotides and immediately available to the cell's metabolism.[12] Studies have not yet appeared on the possible importance of this parameter in platelets or any other secretory cell. More than a decade ago however, Reed[13] showed for rat liver and kidney that the AEC and phosphorylation potential varied in parallel over a broad range of values. Preliminary studies indicate that this also may hold for platelets.[14] How these rather static parameters determine ATP-availability and thereby responsivity in a secretory cell is, however, far from clear.

III. HIGH RESPONSIVITY CORRESPONDS TO A HIGH BASAL ENERGY CONSUMPTION

All cells require a minimal amount of energy consumption for the maintenance of their integrity, for example in ion, membrane, and protein homeostasis. This minimal energy requirement may be expected to be similar for secretory and nonsecretory cells after proper correction for the different cytosol volumes. However, when resting energy consumption rates are compared for a number of different blood cells, a broad range of values emerges, with a more than 150-fold difference between erythrocytes and platelets (Table 1). Both are anucleate cells that lack a number of energy-consuming processes common to other cells, such as RNA and protein turnover, biosynthesis of complex carbohydrates etc. The high ATP-turnover in resting platelets may therefore be related to the specific functions of these cells, aggregation and secretion of a variety of granule-stored substances (see Volume II, Chapter 16). In fact, a few years ago it was suggested that the energy requirement of platelet responses originated *entirely* at the prestimulatory stage,[25] but this view is now abandoned.[8] Basal energy consumption in neutrophils is comparable to that in platelets, while that in lymphocytes is somewhat less (Table 1). Secretory responses of neutrophils are as rapid as those in platelets, whereas the reponses of lymphocytes have considerably slower kinetics. A high basal energy consumption may therefore reflect the state of high responsivity.

IV. ACCOUNTING FOR BASAL ENERGY CONSUMPTION

Identification and quantification of the various ATP-consuming processes in the resting cell will finally establish whether metabolic energy is essential for maintaining the cell in a responsive state. Unfortunately, we are only beginning to understand how energy is spent in a resting cell. In mature erythrocytes, the cell with the least complexity, half of the basal energy consumption is accounted for by Na^+,K^+-ATPase (25%), phosphoinositide turnover (20%), phosphoprotein turnover (13%), and Ca^{2+}-ATPase (1%).[18,26-28] Of basal energy consumption in reticulocytes, protein synthesis, Na^+,K^+-ATPase, ATP-dependent proteolysis, and Ca^{2+}-ATPase account for 30, 25, 15, and 3.5%, respectively.[21] No systematic attempt is made to account for the high basal energy consumption in a secretory cell. In platelets, actin treadmilling has been identified as a major energy-consuming process, accounting for 30 to 50% of basal energy consumption.[29,30] Energy consumption of phosphoinositide turnover has been estimated to be about 8% of the basal energy consumption,[31] whereas ATP-consumption in cAMP-turnover accounts for about 10%.[32] These latter processes are likely candidates for energy-requiring processes that are involved in maintaining platelets in a highly responsive but dormant state. Accounting for all basal energy consumption in a resting cell is, however, a highly unattractive approach due to the diversity of cellular ATP-consuming processes.

Table 1

BASAL ENERGY CONSUMPTION RATES IN SOME BLOOD CELLS

Cell type	Species	Cytosol volume (fℓ)	Basal energy consumption rates		Relative rate per volume[d]	Ref.
			Per cell (μmol ATP eq/min/10^10 cells)	Per cell volume (μmol ATP eq/min/mℓ cells)		
Erythrocytes	Rabbit	—	—	0.033	(1)	18, 19
	Human	90[a]	0.04	0.045	1.5	20
Reticulocytes	Rabbit	—	—	2.25	68	19, 21
Lymphocytes	Human	110[b]	3.3	3.0	90	22
Neutrophils	Human	300[b]	13	4.3	130	23
Platelets	Human	7[c]	0.35—0.70	5—10	150—300	4—6, 24

[a] Taken from Reference 15.
[b] Roughly estimated from data presented in Reference 16
[c] Taken from Reference 17.
[d] Basal energy consumption rates per cell volume were calculated relative to that of rabbit erythrocytes (taken as 1).

Table 2
SOME CONDITIONS THAT MUST BE MET PRIOR TO STIMULATION TO ALLOW FOR AN OPTIMAL STIMULUS-RESPONSE COUPLING

After stimulation	Before stimulation
Receptor occupancy	Receptor exposure
	Maintenance of agonist binding properties
Transmembrane signaling	Availability of GTP
	Prevention of unwanted coupling
	Maintenance of membrane fluidity
Generation of second messengers	
cAMP system	Basal cAMP homeostasis
	Availability of substrate ATP
Calcium/DAG system	Basal $[Ca^{2+}]_i$ homeostasis
	Phosphatidylinositols in a high phosphorylated state
	Maintenance of intracellular Ca^{2+}-store(s)
Effect of second messengers	
Activation of protein kinases	Phosphoprotein homeostasis
Cell morphology changes and cell contractility	Active actin and tubulin state
Extrusion of granule material	Synthesis and storage of granule material
	Maintenance of granule integrity

V. PROCESSES THAT MAY BE INVOLVED IN MAINTAINING RESPONSIVITY

Along the sequence of stimulus response coupling several steps occur that require a certain initial state in order to allow optimal efficiency. Some conditions must be met in the unstimulated cell to enable a secretagogue to initiate this sequence, conditions that are maintained at the expense of metabolic energy (Table 2). Obviously, to allow for a rapid secretory response, the material that is secreted must have been synthesized and stored in granules well before the actual stimulation event, which may explain some of the high basal metabolic activity of secretory cells (see Chapters 3 and 4). It must be emphasized, however, that this cannot explain the high basal energy consumption of platelets, where these functions are virtually absent. Apparently, other energy-consuming steps are responsible, and some of the possibilities will be dealt with here in more detail.

A. Receptor Status

For the intracellularly located steroid receptors, it has long been recognized that their ability to bind ligand depends on the availability of ATP in the cell. As early as 1968, Munck and Brinck-Johnsen[33] found that depleting rat thymus lymphocytes of ATP resulted in the loss of glucocorticoid binding activity, which was reversed upon restoration of cellular ATP. It is now established that the glucocorticoid receptor is a phosphoprotein,[34] and that its phosphorylation state regulates steroid binding,[35,36] although a direct facilitating effect of ATP cannot be excluded.[37] The role of phosphorylation in the regulation of binding activity is also reported for estrogen[38] and progesterone[39] receptors.

More recently, an energy dependency of agonist binding was also described for some cell-surface receptors. Depletion of cellular ATP leads to reduced insulin binding to rat soleus muscle cells,[40,41] adipocytes,[42] and hepatocytes,[43] independent of the inhibition of receptor internalization and replenishment. Yu and Gould[40] supposed therefore, that insulin binding is also regulated by phosphorylation of the receptor molecule, or some closely associated membrane protein. Interestingly, the insulin receptor itself is a protein kinase that is activated by insulin and phosphorylates itself.[44] In contrast, ATP depletion in human lymphocytes results in an *increased* insulin binding, probably through an increase in binding affinity.[45] Also here, phosphorylation steps may be involved but a direct inhibitory effect of ATP on insulin binding to isolated receptor molecules has also been reported.[46] The exposure of prolactin receptors on rat mammary tumors shows an energy-dependency too, because a depletion of cellular ATP coincides with a marked increase in the number of specific binding sites;[47] in this case, extra binding sites are mobilized from a pool of inactive receptor molecules already present in the plasma membrane.

Unfortunately, little is known about the relation between receptor state and energy state in secretory cells. An important role of phosphorylation/dephosphorylation in regulating receptor activity is, however, strongly indicated. Phosphorylation of adenylate cyclase-coupled β-adrenergic receptors[48] and of inositol phospholipid cycle-coupled α_1-adrenergic receptors[49] that correlated with alterations in ligand binding, has been described. It may therefore be expected that a cell's receptor state is dependent on its energy status. Very recently, exposure of the fibrinogen-binding site on platelet-activating factor (PAF)-activated, washed platelets was found to be inhibited by cellular ATP-depletion;[50] occupation of this binding site is essential for the induction of secretory responses by weak agonists such as PAF. In the unstimulated platelet, the binding molecules are already present in the plasma membrane, but in a cryptic state;[51,52] upon stimulation, they are activated, a process which now appears energy dependent.

B. Coupling Factors

GTP plays a crucial role in linking receptor-occupancy to the generation of intracellular second messengers. GTP-regulatory proteins are identified for the receptor-coupled stimulation (N_s) and inhibition (N_i) of adenylate cyclase.[53]

Recently, a strong potentiating effect of GTP was also found for Ca^{2+}-mediated secretion in permeabilized platelets,[54] mast cells,[55] and adrenal cells;[56] both the activity and the Ca^{2+}-sensitivity of polyphosphoinositide phosphodiesterase are markedly increased by GTP[57,58] and GTP-regulatory proteins are therefore, also thought to mediate receptor-coupled activation of the polyphosphoinositol cycle.[59] For N_s and N_i it is known that receptor occupancy results in the binding of GTP, which enables coupling to adenylate cyclase.[53] To allow for optimal coupling GTP must be available at the moment of cell stimulation. N_s is shown to have a significant basal GTPase activity which is proposed to keep the GTP-regulatory protein, and thereby adenylate cyclase, in the resting state.[60] Hence, GTP turnover may be a necessary feature for the resting secretory cell. Little is known of the metabolism of GTP. Cytoplasmic GTP is generally thought to be in close equilibrium with cytoplasmic ATP and hence, to reflect cellular energy status. Under some conditions, however, cytoplasmic ATP in platelets can be markedly reduced while GTP remains essentially unaffected;[61] this may be due to a slow nucleoside diphosphate kinase reaction, or to a different compartmentalization.

GTP-induced activation of adenylate cyclase through N_s is strongly influenced by the fluidity of the cell's plasma membrane, with optimal coupling only occurring over a narrow traject of membrane fluidity states.[62] Maintenance of the optimal fluidity state of intact erythrocyte membrane is also affected by cellular energy status.[63]

C. cAMP Turnover

Basal levels of cellular cAMP are normally in the (sub)micromolar range. In all cell types, several proteins are rapidly labeled when the cells are incubated with ^{32}P-inorganic phosphate, which indicates that phosphoproteins are constantly cycled through phosphorylation-dephosphorylation processes, even under resting conditions. For S49 lymphoma cells, some of these proteins are identified as substrates of cAMP-dependent protein kinases,[64] which shows that these kinases are already active at basal cAMP-levels.

Recent studies onto the kinetics of ^{18}O-incorporation into nucleotide α-phosphoryl groups in intact cells have revealed that, in addition to the phosphorylation state of proteins, cellular cyclic nucleotide metabolism is also highly dynamic.[32,65] In resting platelets cAMP turns over with a t_{y_2} of 200 msec, thereby accounting for approximately 10% of the cell's total energy consumption at rest.[32] Goldberg et al.[65] suggested that in some processes responsivity may be determined by the turnover rate of cyclic nucleotides, rather than by actual concentrations; how a cell can monitor flux through this system is however, unclear. The free energy of phosphodiesteratic cleavage of cyclic nucleotides has been estimated at 11 to 12 kcal/mol;[66] this may be utilized by the cell in making a process thermodynamically favorable by coupling it to the phosphodiesterase reaction. An important role may therefore be indicated for such phosphodiesterase-governed processes. In some cells, such as platelets and neutrophils, secretory responses are inhibited by an elevation of cAMP. In platelets, one of the targets of cAMP-dependent protein kinase has been identified as a Ca^{2+}-ATPase, located in the "dense-tubular system", which is the intracellular Ca^{2+}-store in these cells.[67] This cAMP-dependent ATPase transports Ca^{2+} from the cytoplasm into the lumen of this membrane system, and hence, contributes to maintain low basal $[Ca^{2+}]_i$ and simultaneously builds up the Ca^{2+}-store that is apparently mobilized upon platelet activation. Basal cAMP-metabolism may therefore, be involved in keeping the cells in a resting state. In cells that are activated through an increase in cAMP such a role is not evident, but here a role in maintaining sensitivity may be proposed (see below).

D. Phosphoinositide Turnover

Inositol lipids are implicated in coupling receptor occupancy to the generation of DAG and intracellular calcium. This is indicated by the loss of responsivity of blowfly salivary glands to serotonin[68] and the failure of regenerating rat liver to develop responsivity to angiotensin or vasopressin,[69] when deprived of inositol. Of all inositol lipids, 10 to 20% is present in the higher phosphorylated forms PIP and PIP_2. When cells are incubated with ^{32}P-P_i the monoester-phosphate groups of these minor lipids are rapidly and heavily labeled, in contrast to the diester phosphate.[30] This suggests a high turnover between PI, PIP, and PIP_2 through PI(P) kinases and PI(P) monoesterases. This is further supported by the observation that at all times during incorporation of ^{32}P-P_i into platelets, the specific radioactivity of the monoester-phosphates in PIP_2 is almost identical to that of the terminal ATP-phosphates.[31] ATP consumption in this phosphoinositide turnover is assessed at 8% of total energy consumption in platelets at rest.[31]

At present it is generally assumed that the phosphodiesteratic cleavage of PIP_2 is the initial receptor-controlled step in the induction of Ca^{2+}-mediated responses.[70,71] Again in platelets, the supply of PIP_2 from PI is not, or only slightly accelerated upon activation with thrombin.[31,71] Maintaining the inositol lipids in a highly phosphorylated state seems therefore, vital for a secretory cell. It should be noted, however, that in erythrocytes phosphoinositide turnover contributes to an even higher extent to total energy consumption, as measured by the German group.[27,28] In contrast, another laboratory recently reported a much lower value of ATP turnover in erythrocyte phosphoinositide metabolism.[72] According to this report, less than 1% of basal energy consumption is accounted for by this process. The earlier estimate by the German group of more than 20% might have been caused by improper handling of the erythrocytes.[72]

E. Ca^{2+}-Homeostasis

In quiescent cells, the concentration of intracellular free, ionized calcium ($[Ca^{2+}]_i$) is stable in the range 10 to 100 nM. This is more than 10^4 times lower than in the extracellular fluid, and up to 10^6 times lower than in the intracellular Ca^{2+}-stores, the endoplasmic reticulum (ER), and the mitochondria. These steep gradients across the membranes are maintained by the active extrusion of Ca^{2+} from the cytoplasm at the expense of a continuous consumption of metabolic energy. Ca^{2+}-pumping is most commonly supported by (Ca^{2+}, Mg^{2+})-ATPase or Na^{+}/Ca^{2+}-exchange activities; the latter is considered of minor importance in resting cell Ca^{2+}-homeostasis.[73,74] In addition, uptake of Ca^{2+} by mitochondria is not favored at basal $[Ca^{2+}]_i$, but may play a role at elevated $[Ca^{2+}]_i$.[73,75] In intact cells, $[Ca^{2+}]_i$ rapidly increases 2.5-fold upon depletion of cellular ATP,[76,77] whereas in permeabilized cells omission of ATP results in 3-fold higher steady states for $[Ca^{2+}]_i$.[78]

In platelets, this increase in $[Ca^{2+}]_i$ is accompanied by a decrease in the ER-Ca^{2+}-pool.[76] Besides maintaining $[Ca^{2+}]_i$ at a low level, Ca^{2+}-homeostasis in resting cells also aims at the proper storage of Ca^{2+} for subsequent release during stimulation. Induction of secretion in almost all cells in which calcium mediates the responses depends on the mobilization of Ca^{2+} from intracellular stores. This notion is strongly supported by recent evidence that inositol 1,4,5-trisphosphate (IP$_3$) is able to mobilize calcium from intracellular, nonmitochondrial stores in a variety of secretory cells.[70,78] The amount of sequestered calcium available for release into the cytoplasm may affect responsivity. In line with this, prolonged incubation of platelets with EGTA which readily reduces the intracellular Ca^{2+}-store,[76] is deleterious for their responsivity.[79]

F. Actin Treadmilling

Actin is a major constituent of nonmuscle cells. In most unstimulated cells, 10% is present in filaments as F-actin. Immediately following cell activation the actin polymerization state increases in a variety of cells (References 80 and 81, and citations therein). This increased polymerization is thought to be involved in secretory responses and cell morphology changes. In addition, these filaments serve as a support for glyco(geno)lysis; several glyco(geno)lytic enzymes bind to F-actin, which markedly affects their catalytic properties.[82,83] Hence, formation of filaments will lead to an acceleration of glyco(geno)lytic flux resulting in an enhanced production of ATP adjacent to the sites of its utilization.

Formation of new filaments from monomeric actin is slow, nucleation being the rate-limiting step,[84] which seems incompatible with the observed rapid changes in cell morphology during secretion responses. During the acrosomal reaction of invertebrate sperm, rates of actin polymerization as high as 10 μm/sec are achieved within seconds after stimulation.[85] It is more likely therefore, that actin polymerization occurs through elongation of pre-existing microfilaments. A steady state where a large pool of monomers co-exists with a number of short polymers would allow for the rapid responses seen in secretory cells. Elongation occurs by the successive addition of actin-ATP complexes;[84] the nucleotide is dephosphorylated in this process, but remains bound to the filament. Working with unstimulated platelets, Daniel et al.[86,87] found that this actin-bound ADP, though not immediately available to cellular energy metabolism, is in close equilibrium with cytoplasmic ATP. This was interpreted to indicate a rapid exchange between cytoplasmic and actin-bound nucleotides through actin-treadmilling, the simultaneous addition and release of actin residues at opposite ends of the filament. Based on the difference in incorporation of [^{3}H] adenine into the cytosolic ATP and ADP bound to F-actin in intact platelets, these authors determined the rate of actin treadmilling at 1.8 μmol actin-ATP complexes per minute per 10^{11} cells, compared to a total platelet-actin content of 0.8 μmol/10^{11} cells.[29,30] Hence, ATP-turnover in actin-treadmilling represents the major energy-consuming process in these cells, accounting for 30 to 50% of basal energy consumption. Inhibition of actin-treadmilling by cytochalasin led to a 50% reduction of basal energy consumption, resulting in a similar estimate for the contribution of this process.[6] It is tempting to speculate that the rapid and extensive redistribution in actin seen in platelets upon receptor-mediated activation,[88] is made possible by this high cycling in the unstimulated cell. Nothing is known, however, on the dynamics of actin in other secretory cells at rest.

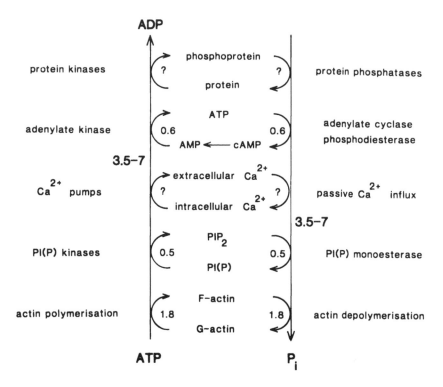

FIGURE 1.   Some circular pathways in unstimulated platelets. A number of circular pathways account for more than half of basal energy consumption in platelets (see text). In the steady state the levels of the intermediates are constant but turn over rapidly. For thermodynamic reasons, a continuous input of metabolic energy, mainly as supplied by ATP, is required to keep two opposing pathways operative at the same time. Numbers give approximate fluxes, in μmol of ATP equivalents/min/10^{11} cells.

VI. MAINTENANCE OF SENSITIVITY

It is remarkable that a quiescent secretory cell, such as the platelet, maintains such a highly dynamic steady state, where maintenance of constant levels of cAMP, $[Ca^{2+}]_i$, PIP_2, F-actin, etc. seems to be sufficient to preserve responsivity (Figure 1). This dynamic state is very expensive for the cell in terms of energy, and therefore, rather seems harmful. However, similar cycling is frequently encountered in metabolic pathways, and seems to be a very common feature of cellular metabolism. In the last decade Newsholme and colleagues have developed the hypothesis that this cycling serves in the control of sensitivity of the pathways in which they are integrated.[89-91] These workers therefore, adopted the term "substrate cycle" to stress that this cycling is, by no means, futile.

A. Substrate Cycling and Gain Control

A substrate cycle is composed of two opposing reactions, at least one of which is nonequilibrium, and both are operative at the same time.[91] Figure 2 depicts a pathway in which a substrate cycle is present between the intermediates A and B. In the steady state, the net flux from A to B (F-C) equals the flux through the whole pathway (J). Alteration of J can be produced by changes in B through effectors of either the forward flux or the reverse flux of the substrate cycle, or both. The advantage of having two distinct opposing steps is that both forward and reverse steps can be modulated independently, thereby offering the potential for gain control. A role for substrate cycles is proposed for both mechanisms of sensitivity and magnitude amplification.[92] In sensitivity amplification, a percentage change

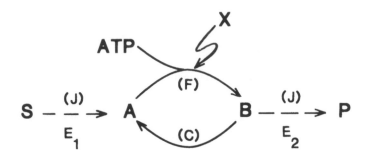

FIGURE 2. A substrate cycle in a simple pathway. A steady flux (J) is flowing through the sequence from S to P; F and C are the fluxes through the forward and reverse pathways, respectively, of the cycle that connects A and B. The cycle improves the sensitivity of the flux J to changes in effector X provided that the elasticity coefficient of E_2 to B exceeds that of the reverse reaction to B.[90] The magnitude of this improvement depends on the relative cycling rate (C/J), the elasticity coefficient of E_1 to A, and the elasticity coefficients of E_2, F, and C to their respective substrates.[90,95] The substrate cycle can maximally improve sensitivity of the sequence to changes in X by a factor $1 + (C/J)$.

in stimulus (input signal) produces a larger percentage change in output response. This can be achieved by a change in the concentration of effectors that increases F and simultaneously reduces C. Under some conditions, substrate cycles may produce switch on-switch off effects, for example in zero-order ultrasensitivity, where the reverse reaction is saturated for B,[93] or in some cases of interconvertible enzymes.[94] In magnitude amplification, an input signal produces a larger, but linearly proportionate output response. Here, substrate cycles can protect the cell against too excessive signal amplification, when an increase in B is counteracted by an acceleration of C through mass-action effects on the reverse reaction.[92]

It must be emphasized that these properties for gain control would not be possible if there were a simple equilibrium between A and B. In order to maintain two opposing reactions operative at the same time, a continuous input of energy, notably as supplied by ATP, is necessary. This consumption of energy is, in fact, the price the cell must pay for an increased sensitivity to environmental changes.

B. Cycling Rate and Sensitivity

The maximum factor with which a substrate cycle can increase the sensitivity of a pathway to effectors of the forward reaction is derived by Newsholme and Crabtree as $1 + (C/J)$, in which C denotes the rate of the B to A reaction, and hence, the cycling rate.[91] The actual sensitivity also depends on the properties of the interactions between cycle intermediates and the adjacent components of the particular pathway, which in general will result in a somewhat lower factor.[90,95] It follows from this model that a substrate cycle can improve sensitivity if the cycling rate (C) is substantially higher than the net flux (J). This condition is surely met in the resting platelet (Figure 1), where net flux through all the mentioned cycles is negligible whereas turnover between the cycle intermediates is considerable. However, an exception should be made for the phosphoinositide turnover, since here the effector (receptor occupancy) is thought to activate a step outside the cycle itself.[70,71] Instead, it may be considered as a PIP_2-branchpoint, of which one branch (the phosphodiesterase) is negligible compared to the second (the phosphomonoesterase) in the resting state (Figure 3). As pointed out previously,[90,95] this situation is particularly suited to improve sensitivity of the phosphodiesterase branch.

Another prediction that can be made from the model is that effectors that stimulate both the forward and the reverse reaction of the cycle in parallel, can increase sensitivity of the

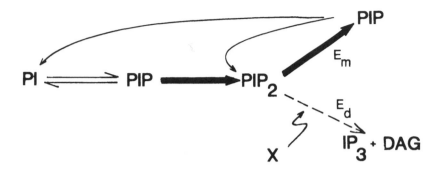

FIGURE 3. The PIP_2 branchpoint. In unstimulated cells, PIP_2 (phosphatidylinositol 4,5-bisphosphate) is continuously dephosphorylated through a phosphomonoesterase (E_m) and rephosphorylated again through a PIP-kinase. A negligible fraction of total flux through PIP_2 might divert to IP_3 and DAG through a phosphodiesterase (E_d), even under resting conditions; this might be supported by the consistent finding of low but significant basal IP_3 levels.[101-103] The sensitivity of E_d to effector X (receptor occupancy) approaches a maximum when the flux through E_m is much greater than the flux through E_d.[90] The benefit of the substrate cycle between PI(P) and PIP_2 might lie in combining flux-switching potential with conservation of the PI(P)-backbone.

pathway.[89] Such an increased cycling would, of course, be accompanied by a higher energy consumption. Recent work of Newsholme's group has shown that the cycling rate of several substrate cycles can be markedly altered.[89] In this respect, the stress and thyroid hormones are of particular interest. For both types it is established that they increase energy expenditure as well as sensitivity of several tissues; both are now also shown to increase the cycling rate of some cycles.[96] The resting secretory cell, exemplified by the platelet, is obviously equipped with similar mechanisms, but at present experimental support is not available.

An increased cycling can enhance the speed with which a new steady-state concentration of B is attained upon addition of another effector that activates F and C disproportionately; the pathway is "primed" for a more rapid response by the increased cycling. This might well be a basis for synergism. Synergism between different agonists is a well-documented phenomenon in almost all secretory cells, though poorly understood. It is commonly interpreted as reflecting the additional effect of subthreshold generation of a second messenger, or the involvement of different second messengers separately generated by the different agonists. An increased turnover of one of the substrate cycles such as summarized in Figure 1, by one agonist can potentiate the responses induced by the second, without changes in any of the putative second messengers. Also here, the stress hormone epinephrine might be of interest. At low concentrations, this agent strongly potentiates platelet activation by thrombin, ADP, and PAF.[97,98] It is assumed that this is achieved through the inhibition of adenylate cyclase, but a reduction in cellular cAMP has never been observed under those conditions.[99] The potentiating effect of epinephrine might well result from a change in cAMP-cycling, or from one of the other cycles.

VII. CONCLUSION

I have attempted to show here that basal energy consumption is high in blood platelets, a feature that may also hold for other secretory cells and that may be related to the maintenance of responsivity. Maintenance of a cell's responsivity apparently depends on an energy-controlled dynamic state of target molecules; a resting cell can be likened to a car: in order to get a good start the motor must be on, the stimulus being just a gear shift. At present, more than 50% of this high basal energy consumption can be accounted for. A number of circular pathways have been identified in the resting platelet, which turn over very rapidly, giving the cell an extremely dynamic steady state. I propose that this high cycling, and hence, high basal energy consumption, is the price for the high responsivity of these cells. This model is based on the resemblance of these cycles with the "substrate cycles" that are abundant in metabolic pathways, and which are thought to serve in sensitivity control. In order to evaluate the validity of this proposal (1) other secretory as well as nonsecretory cells need to be examined for the presence of these or similar highly active cycles; (2) the rate of cycling must be shown susceptible to variation; and (3) alterations in cycling rate must correlate with changes in the cell's sensitivity for its agonists.

Several methods for the quantification of this cycling in intact cells, notably involving double-labeling techniques, have been developed in recent years,[30,32,96,100] which may be useful in testing this model.

ACKNOWLEDGMENTS

I am grateful to Dr. H. Holmsen and Dr. O.-B. Tysnes for critically reading the manuscript, and Dr. C. Cook for her attempts to improve the English. The author was a recipient of an NTNF (Royal Norwegian Council for Scientific and Industrial Research) Postdoctoral Fellowship.

REFERENCES

1. **Conn, P. M.,** *Cellular Regulation of Secretion and Release,* Academic Press, New York, 1982, xvii.
2. **Knight, D. E. and Baker, P. F.,** Calcium-dependence of catecholamine release from bovine adrenal medullary cells after exposure to intense electric field, *J. Membr. Biol.,* 68, 107, 1982.
3. **Knight, D. E., Niggli, V., and Scrutton, M. C.,** Thrombin and activators of protein kinase C modulate secretory responses of permeabilised human platelets induced by Ca^{++}, *Eur. J. Biochem.,* 143, 437, 1984.
4. **Akkerman, J. W. N., Gorter, G., Schrama, L., and Holmsen, H.,** A novel technique for rapid determination of energy consumption in platelets. Demonstration of different energy consumption associated with three secretory responses, *Biochem. J.,* 210, 145, 1983.
5. **Verhoeven, A. J. M., Mommersteeg, M. E., and Akkerman, J. W. N.,** Quantification of energy consumption in platelets during thrombin-induced aggregation and secretion. Tight coupling between platelet responses and the increment in energy consumption, *Biochem. J.,* 221, 777, 1984.
6. **Verhoeven, A. J. M., Mommersteeg, M. E., and Akkerman, J. W. N.,** Comparative studies on the energetics of platelet responses induced by different agonists, *Biochem. J.,* 236, 879, 1986.
7. **Panten, U., Zielmann, S., Langer, J., Zunkler, B.-J., and Lenzen, S.,** Regulation of insulin secretion by energy metabolism in pancreatic B-cell mitochondria. Studies with a non-metabolizable leucine analogue, *Biochem. J.,* 219, 189, 1984.
8. **Verhoeven, A. J. M., Mommersteeg, M. E., and Akkerman, J. W. N.,** Metabolic energy is required in human platelets at any stage during optical aggregation and secretion, *Biochim. Biophys. Acta,* 800, 242, 1984.
9. **Verhoeven, A. J. M., Mommersteeg, M. E., and Akkerman, J. W. N.,** Balanced contribution of glycolytic and adenylate pool in supply of metabolic energy in platelets, *J. Biol. Chem.,* 260, 2621, 1985.
10. **Atkinson, D. E.,** *Cellular Energy Metabolism and its Regulation,* Academic Press, New York, 1977, 40.
11. **Fromm, H. J.,** Discussion forum: cellular energy control — control by energy charge is an untenable theory, *Trends Biochem. Sci.,* 2, 198, 1977.
12. **Erecinska, M. and Wilson, D. F.,** Homeostatic regulation of cellular energy metabolism, *Trends Biochem. Sci.,* 3, 219, 1978.
13. **Reed, E. B.,** Coordination of adenylate energy charge and phosphorylation state during ischemia and under physiological conditions in rat liver and kidney, *Life Sci.,* 19, 1307, 1976.
14. **Verhoeven, A. J. M., Marszalek, J., and Holmsen, H.,** manuscript in preparation, 1987.

15. **Williams, W. J., Beutler, E., Erslev, A. J., and Lichtman, M. A.,** *Hematology,* 3rd ed., McGraw-Hill, New York, 1983, 9.

16. **Junquiera, L. C., Carneiro, J., and Contopoulos, A.,** *Basic Histology,* 2nd ed., Lange Medical Publ., Los Altos, Calif., 1977, 228.

17. **Holmsen, H. and Robkin, L.,** Effects of antimycin A and 2-deoxyglucose on energy metabolism in washed human platelets, *Thromb. Haemostasis,* 42, 1460, 1980.

18. **Maretzki, D., Reimann, B., Klatt, D., and Rapoport, S.,** A form of $(Ca^{2+} + Mg^{2+})$-ATPase of human red cell membranes with low affinity for Mg-ATP: a hypothesis for its function, *FEBS Lett.,* 111, 269, 1980.

19. **Siems, W., Muller, M., Dumdey, R., Holzhutter, H.-G., Rathmann, J., and Rapoport, S. M.,** Quantification of pathways of glucose utilization and balance of energy metabolism of rabbit reticulocytes, *Eur. J. Biochem.,* 124, 567, 1982.

20. **Keitt, A. S. and Bennett, D. C.,** Pyruvate kinase deficiency and related disorders of red cell glycolysis, *Am. J. Med.,* 41, 762, 1966.

21. **Siems, W., Dubiel, W., Dumdey, R., Muller, M., and Rapoport, S. M.,** Accounting for the ATP-consuming processes in rabbit reticulocytes, *Eur. J. Biochem.,* 139, 101, 1984.

22. **Borregaard, N. and Herlin, T.,** Energy metabolism of human neutrophils during phagocytosis, *J. Clin. Invest.,* 70, 550, 1982.

23. **Roos, D. and Loos, J. A.,** Effect of phytohaemagglutinin on the carbohydrate metabolism of human blood lymphocytes after inhibition of the oxidative phosphorylation, *Exp. Cell Res.,* 77, 121, 1973.

24. **Akkerman, J. W. N. and Holmsen, H.,** Interrelationships among platelet responses: studies on the burst in proton liberation, lactate production and oxygen uptake during platelet aggregation and Ca^{++}-secretion, *Blood,* 57, 956, 1981.

25. **Holmsen, H.,** Platelet energy metabolism in relation to function, in *Platelets and Thrombosis,* Mills, D. C. B. and Pareti, F. I., Eds., Academic Press, New York, 1977, 45.

26. **Rapoport, S., Maretzki, D., and Siems, W.,** Balance and regulation of ATP-producing and consuming reactions, in *Abstractbook 16th FEBS Meeting,* Moscow, 1984, 22.

27. **Maretzki, D., Reimann, B., Klatt, D., and Schwarzer, E.,** Involvement of polyphosphoinositides in the ATP turnover of intact human erythrocytes and in the ATPase activity of purified membranes, *Biomed. Biochim. Acta,* 42, S 72, 1983.

28. **Reimann, B., Klatt, D., Tsamaloukas, A. G., and Maretzki, D.,** Membrane phosphorylation in intact human erythrocytes, *Acta Biol. Med. Germ.,* 40, 487, 1981.

29. **Daniel J. L., Molish, I. R., Salganicoff, L., and Holmsen, H.,** Measurement of the nucleotide exchange rate as a determination of the state of cellular actin, in *Motility in Cell Function,* Pepe, F. A., Sanger, J. W., and Nachmias, V. T., Eds., Academic Press, New York, 1979, 459.

30. **Daniel, J. L., Molish, I. R., Robkin, L., and Holmsen, H.,** Nucleotide exchange between cytosolic ATP and F-actin-bound ADP may be a major energy-utilizing process in unstimulated platelets, *Eur. J. Biochem.,* 156, 677, 1986.

31. **Verhoeven, A. J. M., Tysnes, O.-B., Aarbakke, G. M., Cook, C. A., and Holmsen, H.,** Turnover of the phosphomonoester groups of polyphosphomonoester groups of polyphosphoinositol lipids in unstimulated human platelets, *Eur. J. Biochem.,* 166, 3, 1987.

32. **Walseth, T. F., Gander, J. E., Eide, S. J., Krick, T. P., and Goldberg, N. D.,** ^{18}O-labeling of adenine nucleotide α-phosphoryls in platelets. Contribution by phosphodiesterase-catalyzed hydrolysis of cAMP, *J. Biol. Chem.,* 258, 1544, 1983.

33. **Munck, A. and Brinck-Johnsen, T.,** Specific and nonspecific physicochemical interactions of glucocorticoids with rat thymus cells in vitro, *J. Biol. Chem.,* 243, 5556, 1968.

34. **Singh, V. B. and Moudgil, V. K.,** Phosphorylation of the glucocorticoid receptor, *J. Biol. Chem.,* 260, 3684, 1985.

35. **Nielsen, C. J., Sando, J. J., and Pratt, W. B.,** Evidence that dephosphorylation inactivates the glucocorticoid receptor, *Proc. Natl. Acad. Sci. U.S.A.,* 74, 1398, 1977.

36. **Sando, J. J., LaForest, A. C., and Pratt, W. B.,** ATP-dependent activation of L-cell glucocorticoid receptors to the steroid binding form, *J. Biol. Chem.,* 254, 4772, 1979.

37. **Towle, A. C. and Sze, P. Y.,** Inactivation by Na^+,K^+-ATPase of cytosol glucocorticoid receptors from rat brain and liver, *Mol. Cell. Biochem.,* 52, 145, 1983.

38. **Migliaccio, A., Lastoria, S., Moncharmont, B., Rotondi, A., and Auricchio, F.,** Phosphorylation of calf uterus 17β-estradiol receptor by endogenous Ca^{2+}-stimulated kinase activating the hormone binding of the receptor, *Biochem. Biophys. Res. Commun.,* 109, 1002, 1982.

39. **Puri, R. K., Dougherty, J. J., and Toft, D. O.,** The avian progesterone receptor: isolation and characterization of phosphorylated forms, *J. Steroid Biochem.,* 20, 23, 1984.

40. **Yu, K. T. and Gould, M. K.,** Effect of prolonged anaerobiosis on ^{125}I-insulin binding to rat soleus muscle: permissive effect of ATP, *Am. J. Physiol.,* 235, E606, 1978.

41. **Lim, Y. H. and Gould, M. K.,** ATP-dependence of ^{125}I-insulin binding by rat soleus muscle, *Biochem. Int.*, 6, 163, 1983.

42. **Steinfelder, H. J. and Joost, H. G.,** Reversible reduction of insulin receptor affinity by ATP depletion in rat adipocytes, *Biochem. J.*, 214, 203, 1983.

43. **Draznin, B., Solomons, C. C., Toothaker, D. R., and Sussmann, K. E.,** Energy-dependent steps in insulin-hepatocyte interaction, *Endocrinology*, 108, 8, 1981.

44. **Roth, R. A. and Cassel, D. J.,** Insulin-receptor: evidence that it is a protein kinase, *Science*, 219, 299, 1983.

45. **Vigneri, R., Maddux, B., and Goldfine, I. D.,** The effect of phenformin and other adenosine triphosphate (ATP)-lowering agents on insulin binding to IM-9 cultured lymphocytes, *J. Cell. Biochem.*, 24, 177, 1984.

46. **Trischitta, V., Vigneri, R., Roth, R. A., and Goldfine, I. D.,** ATP and the other nucleoside triphosphates inhibit the binding of insulin to its receptor, *Metabolism*, 33, 577, 1984.

47. **Costlow, M. E. and Hample, A.,** Prolactin receptors in cultured rat mammary tumor cells. Unmasking of cell surface receptors by energy depletion, *J. Biol. Chem.*, 257, 6971, 1982.

48. **Sibley, D. R. and Lefkowitz, R. J.,** Molecular mechanisms of receptor desensitization using the β-adrenergic receptor-coupled adenylate cyclase system as a model, *Nature (London)*, 317, 124, 1985.

49. **Leeb-Lundberg, L. M. F., Cotecchia, S., Lomasney, J. W., DeBernardis, J. F., Lefkowitz, R. J., and Caron, M. G.,** Phorbol esters promote α_1-adrenergic receptor phosphorylation and receptor uncoupling from inositol phospholipid metabolism, *Proc. Natl. Acad. Sci. U.S.A.*, 82, 5651, 1985.

50. **Kloprogge, E., Hasselaar, P., and Akkerman, J. W. N.,** PAF-acether (1-O-hexadecyl/octa decyl-2-acetyl-sn-glycero-3-phophocholine)-induced fibrinogen binding depends on metabolic energy, *Biochem. J.*, 238, 885, 1986.

51. **Shattil, S. J., Hoxie, J. A., Cunningham, M., and Brass, L. F.,** Changes in the platelet membrane glycoprotein IIb-IIIa complex during platelet activation, *J. Biol. Chem.*, 260, 11107, 1985.

52. **Fitzgerald, L. A. and Phillips, D. R.,** Calcium regulation of the platelet membrane glycoprotein IIb-IIIa complex, *J. Biol. Chem.*, 260, 11366, 1985.

53. **Gilman, A. G.,** G-proteins and dual control of adenylate cyclase, *Cell*, 36, 577, 1984.

54. **Haslam, R. J. and Davidson, M. M. L.,** Guanine nucleotides decrease the free [Ca^{2+}] required for secretion of serotonin from permeabilized blood platelets. Evidence of a role for a GTP-binding protein in platelet activation, *FEBS Lett.*, 174, 90, 1984.

55. **Gomperts, B. D.,** Involvement of guanine nucleotide-binding protein in the gating of Ca^{++} by receptors, *Nature (London)*, 306, 64, 1983.

56. **Knight, D. E. and Baker, P. F.,** Guanine nucleotides and Ca-dependent exocytosis. Studies on two adrenal cell preparations, *FEBS Lett.*, 189, 345, 1985.

57. **Uhing, R. J., Jiang, H., Prpic, V., and Exton, J. G.,** Regulation of liver plasma membrane phospho-inositide phosphodisterase by guanine nucleotides and calcium, *FEBS Lett.*, 188, 317, 1985.

58. **Sasaguri, T., Hirata, M., and Kuriyama, H.,** Dependence on Ca^{2+} of the activities of phosphatidylinositol 4,5-bisphosphate phosphodiesterase and inositol 1,4,5-trisphosphate phosphatase in smooth muscles of the porcine coronary artery, *Biochem. J.*, 231, 497, 1985.

59. **Cockcroft, S. and Gomperts, B. D.,** Role of guanine nucleotide binding protein in the activation of polyphosphoinositide phosphodiesterase, *Nature (London)*, 314, 534, 1985.

60. **Rodbell, M.,** Programmable messengers: a new theory of hormone action, *Trends Biochem. Sci.*, 10, 461, 1985.

61. **Holmsen, H.,** Nucleotide metabolism of platelets, *Annu. Rev. Physiol.*, 47, 677, 1985.

62. **Salesse, R., Garnier, J., and Daveloose, D.,** Modulation of adenylate cyclase activity by the physical state of pigeon erythrocyte membrane. II. Fluidity-controlled coupling between the subunits of the adenylate cyclase system, *Biochemistry*, 21, 1587, 1982.

63. **Kamada, T., Setoyama, S., Chuman, Y., and Otsuji, S.,** Metabolic dependence of the fluidity of intact erythrocyte membrane, *Biochem. Biophys. Res. Commun.*, 116, 547, 1983.

64. **Steinberg, R. A. and Kiss, Z.,** Basal phosphorylation of cyclic AMP-regulated phosphoprotein in intact S49 mouse lymphoma cells, *Biochem. J.*, 227, 987, 1985.

65. **Goldberg, N. D., Ames, A., III, Gander, E., and Walseth, T. F.,** Magnitude of increase in retinal cGMP metabolic flux determined by ^{18}O incorporation into nucleotide α-phosphoryls corresponding with intensity of photic stimulation, *J. Biol. Chem.*, 258, 9213, 1983.

66. **Greengard, P., Hayaishi, O., and Colowick, S. P.,** Enzymatic adenylylation of pyrophosphate by 3′,5′ cyclic AMP; reversal of the adenyl cyclase reaction, *Fed. Proc.*, 28, 467, 1969.

67. **Käser-Glanzmann, R., Gerber, E., and Lüscher, E. F.,** Regulation of the intracellular calcium level in human blood platelets: cyclic adenosine 3′,5′-monophosphate dependent phosphorylation of a 22,000 dalton component in isolated Ca^{++}-accumulating vesicles, *Biochim. Biophys. Acta*, 558, 334, 1979.

68. **Berridge, M. D. and Fain J. N.**, Inhibition of phosphatidylinositol synthesis and the inactivation of calcium entry after prolonged exposure of the blowfly salivary gland to 5-hydroxytryptamine, *Biochem. J.*, 178, 59, 1979.

69. **Huerta-Bahena, J. and Garcia-Sainz, J. A.**, Inositol administration restores the sensitivity of liver cells formed during liver regeneration to alpha$_1$-adrenergic amines, vasopressin and angiotensin II, *Biochim. Biophys. Acta*, 763, 125, 1983.

70. **Berridge, M. D.**, The molecular basis of communication within the cell, *Sci. Am.*, 253, 124, 1985.

71. **Wilson, D. B., Neufeld, E. J., and Majerus, P. W.**, Phosphoinositide interconversion in thrombin-stimulated human platelets, *J. Biol. Chem.*, 260, 1046, 1985.

72. **Dale, G. L.**, Quantitation of adenosine-5'-triphosphate used for phosphoinositide metabolism in human erythrocytes, *Blood*, 66, 1133, 1985.

73. **Carafoli, E., and Penniston, J. T.**, The calcium signal, *Sci. Am.*, 253, 50, 1985.

74. **Brass, L. F.**, The effect of Na$^+$ on Ca^{2+}-homeostasis in unstimulated platelets, *J. Biol. Chem.*, 259, 12571, 1984.

75. **Somlyo, A. P.**, Cellular site of calcium regulation, *Nature (London)*, 309, 517, 1985.

76. **Brass, L. F.**, Ca^{2+}-homeostasis in unstimulated platelets, *J. Biol. Chem.*, 259, 12563, 1984.

77. **Snowdowne, K. W., Freudenrich, G. C., and Borle, A. B.**, The effects of anoxia on cytosolic free calcium, calcium fluxes, and cellular ATP-levels in cultured kidney cells, *J. Biol. Chem.*, 260, 11619, 1985.

78. **Prentki, M., Wollheim, C. B., and Lew, P. D.**, Ca^{2+}-homeostasis in permeabilized human neutrophils. Characterization of Ca^{2+}-sequestering pools and the action of inositol 1,4,5-trisphosphate, *J. Biol. Chem.*, 259, 13777, 1984.

79. **Gerrard, J. M., Peterson, D. A., and White, J. G.**, Calcium mobilization, in *Platelets in Biology and Pathology*, Vol. 2, Gordon, J. L., Ed., Elsevier/North-Holland, Amsterdam, 1981, 407.

80. **Rao, K. M. K., Betschart, J. M., and Virji, M. A.**, Hormone-induced actin polymerization in rat hepatoma cells and human leucocytes, *Biochem. J.*, 230, 709, 1985.

81. **Sklar, L. A., Omann, G. M., and Painter, R. G.**, Relationship of actin polymerization and depolymerization to light scattering in human neutrophils: dependence on receptor occupancy and intracellular Ca^{++}, *J. Cell Biol.*, 101, 1161, 1985.

82. **Poglazov, B. F.**, Actin and coordination of metabolic processes, *Biochem. Int.*, 6, 757, 1983.

83. **Masters, C. J.**, Interactions between soluble enzymes and subcellular structure, *CRC Crit. Rev. Biochem.*, 11, 105, 1981.

84. **Pollard, T. D. and Craig, S. W.**, Mechanism of actin polymerization, *Trends Biochem. Sci.*, 7, 55, 1982.

85. **Tilney, L. G., Hatano, S., Ishikawa, H., and Mooseker, M. S.**, The polymerization of actin: its role in the generation of the acrosomal process of certain echinoderm sperm, *J. Cell Biol.*, 59, 109, 1977.

86. **Daniel, J. L., Robkin, L., Molish, I. R., and Holmsen, H.**, Determination of the ADP concentration available to participate in energy metabolism in an actin-rich cell, the platelet, *J. Biol. Chem.*, 254, 7870, 1979.

87. **Daniel, J. L., Molish, I. R., and Holmsen, H.**, Radioactive labeling of the adenine nucleotide pool of cells as a method to distinguish among intracellular compartments. Studies on human platelets, *Biochim. Biophys. Acta*, 632, 444, 1980.

88. **Fox, J. E. B. and Phillips, D. R.**, Polymerization and organization of actin filaments within platelets, *Semin. Hematol.*, 20, 243, 1983.

89. **Newsholme, E. A., Challiss, R. A. J., and Crabtree, B.**, Substrate cycles: their role in improving sensitivity in metabolic control, *Trends Biochem. Sci.*, 9, 277, 1984.

90. **Crabtree, B. and Newsholme, E. A.**, A quantitative approach to metabolic control, *Curr. Top. Cell. Regul.*, 25, 21, 1985.

91. **Newsholme, E. A. and Crabtree, B.**, Substrate cycles in metabolic regulation and in heat generation, *Biochem. Soc. Symp.*, 41, 61, 1976.

92. **Koshland, D. E., Jr., Goldbeter, A., and Stock, J. B.**, Amplification and adaptation in regulatory and sensory systems, *Science*, 217, 220, 1982.

93. **LaPorte, D. C. and Koshland, D. E., Jr.**, Phosphorylation of isocitrate dehydrogenase as a demonstration of enhanced sensitivity in covalent regulation, *Nature (London)*, 305, 286, 1983.

94. **Crabtree, B.**, A metabolic switch produced by enzymically interconvertible forms of an enzyme, *FEBS Lett.*, 187, 193, 1985.

95. **Fell, D. A. and Sauro, H. M.**, Metabolic control and its analysis. Additional relationships between elasticities and control coefficients, *Eur. J. Biochem.*, 148, 555, 1985.

96. **Challiss, R. A. J., Arch, J. R. S., and Newsholme, E. A.**, The rate of substrate cycling between fructose 6-phosphate and fructose 1,6-bisphosphate in skeletal muscle from cold-exposed, hyperthyroid or acutely exercised rats, *Biochem. J.*, 231, 217, 1985.

97. **Steen, V. and Holmsen, H.,** Synergism between thrombin and epinephrine in human platelets: different dose-response relationships for aggregation and dense granule secretion, *Thromb. Haemostasis,* 54, 680, 1985.

98. **Cameron, H. A. and Ardlie, N. G.,** The facilitating effects of adrenaline on platelet aggregation, *Prostaglandins Leukotrienes Med.,* 9, 117, 1982.

99. **Haslam, R. J.,** Roles of cyclic nucleotides in platelet function, *Ciba Found. Symp.,* 35, 121, 1975.

100. **Challiss, R. A. J., Arch, J. R. S., and Newsholme, E. A.,** The rate of substrate cycling between fructose 6-phosphate and fructose 1,6-bisphosphate in skeletal muscle, *Biochem. J.,* 221, 153, 1984.

101. **Rittenhouse, S. E. and Sasson, J. P.,** Mass changes in myoinositol trisphosphate in human platelets stimulated by thrombin. Inhibitory effects of phorbol ester, *J. Biol. Chem.,* 260, 8657, 1985.

102. **Watson, S. P, Reep, B., McConnell, R. T., and Lapetina, E. G.,** Collagen stimulates [^{3}H] inositol trisphosphate formation in indomethacin treated human platelets, *Biochem. J.,* 226, 831, 1985.

103. **Siess, W. and Binder, H.,** Thrombin induces the rapid formation of inositol bisphosphate and inositol trisphosphate in human platelets, *FEBS Lett.,* 180, 107, 1985.

Energy-Dependent Steps in Signal Processing

Chapter 6

LIGAND-INDUCED SURFACE RECEPTOR MOVEMENT AND TRANSMEMBRANE INTERACTIONS

Lilly Y. W. Bourguignon

TABLE OF CONTENTS

I. INTRODUCTION

The plasma membrane constitutes the cell's limiting boundary. Critical functions of the plasma membrane include maintenance and modulation of electrochemical gradients between the cell exterior and the cytoplasm; concentration of membrane bound enzymes; assurance of close topological associations between certain membrane-associated molecules (receptors); and recognition of external signals and their transduction to the cell interior.

Receptors typically bind specific ligands with affinities on the order of 10^{-8} M and they convert that interaction into a sequence of signals that ultimately leads to biological activity. For instance, the β-adrenergic receptors on a pineal cell bind noradrenaline and translate that interaction into an elevated cellular cAMP (cyclic adenosine monophosphate) level. This elevation of cAMP leads to the secretion of melatonin which, in turn, acts on the sex glands. Interactions between lymphocyte surface receptors and various foreign ligands also initiate a signal that travels from the plasma membrane to the cell nucleus which subsequently induces cell proliferation, differentiation, and antibody secretion.

Signal transduction leading from receptor-ligand binding to any particular type of cellular activation appears to involve a number of biological steps. The information available with regard to the molecular mechanisms involved in signal transduction is quite limited at the present time. One possible mechanism proposes that a ligand-induced change in the receptor conformation is required for subsequent biological reactions to occur. For example, the binding of acetylcholine to its postsynaptic membrane receptor induces a new conformation of the receptor which, in turn, leads to a cascade of regulatory events that appear to be important for many neurological responses.[1] Furthermore, kinetic studies on hormone-receptor dissociation (e.g., insulin dissociation from its cell surface receptor) have indicated that the binding mechanism is negatively cooperative.[2] Conformational changes in the insulin receptor may be responsible for these alterations of its binding characteristics. A complete understanding of these ligand-receptor interactions requires a detailed biophysical analysis which to date has not been achieved.

Another possible mechanism for signal transduction at the cell surface is ligand-induced receptor clustering and aggregation (so-called patching and capping, respectively). It has been suggested that receptor patching and capping may be responsible for (1) the mitogenesis of antibody-producing cells;[4-5] (2) the degranulation of mast cells and basophils;[6-7] (3) insulin-dependent glucose oxidation;[8] (4) cellular activation by epidermal growth factor;[9] and (5) toxin-mediated cellular responses.[10] In view of the importance of the physiological effects which may be due to receptor patching and capping, a great deal of effort has been expended in determining the organization of the plasma membrane-associated receptors as well as the regulatory mechanisms involved in the redistribution of the receptors. This paper will focus on the recent advances in our understanding of ligand-induced surface receptor movement.

II. DESCRIPTION OF LIGAND-INDUCED SURFACE RECEPTOR MOVEMENT

Binding of external ligands (e.g., antibodies, lectins, hormones and toxins, etc.) to their specific receptors, generally leads to a rearrangement of the plasma membrane constituents from a uniform distribution (Figure 1a) into patches (clusters) (Figure 1b and 1d) and then caps (aggregates) (Figure 1c and 1d). Patching and capping have been observed with lymphocytes,[11-12] platelets,[13] fibroblasts,[14-15] muscle cells,[16] Ehrlich Ascites tumor cells,[17-18] nerve cells,[19-20] granulosa cells,[21] and cellular slime mold cells.[22] There appears to be an association between receptor capping and certain cellular activities. Indirect evidence for this is suggested by the observations that reduced capping has been found with lymphocytes obtained from patients with certain diseases such as lymphoproliferative disorders, muscular dystrophies, and autoimmune diseases.[24-29] Following patch and cap formation, the ligand-receptor complexes can be internalized by the cell through endocytosis and/or shed into the extracellular environment or medium.[23] A direct correlation may be drawn between the internalization of ligand-receptor complexes and the subsequent replenishment of new receptors on the cell surface.

The binding of specific ligands to cell surface receptors induces a complex sequence of events in the target cell. While these events may vary somewhat with different ligands (e.g., physiologically and nonphysiologically relevant molecules), receptor patching, capping, endocytosis, and intracellular ligand processing occur with all of the ligands studied. Conventional multivalent ligands (e.g., antibodies against surface antigens or lectins) and nonconventional multivalent ligands (e.g., polycationized ferritin and polylysine) both induce only small clusters (so-called patches) at cold temperatures (0 to 4°C). These patches can subsequently form large polarized aggregates (so-called caps) via an energy-dependent process upon incubation at room temperature or 37°C.[11-12]

At this time, our knowledge concerning ligand-surface receptor interactions, and the subsequent events leading to the activation of the cellular responses is still rather meager. It has been suggested that the ligand and its receptor are clustered by a cross-linking event at the cell surface that is highly dependent on the valency of the ligand. In lymphocytes, for example, divalent polyclonal antibodies against surface molecules such as immunoglobulin (Ig) (a major B-lymphocyte specific glycoprotein) and T-200 (a major T-lymphocyte specific glycoprotein) are required for the induction of Ig and T-200 capping, respectively;[11,30] monovalent or monoclonal antibody is unable to cause redistribution of the Ig and T-200 receptors.[12,31] However, in the case of histocompatibility antigens (H-2 in mouse and HLA in man) as well as TL and Thy-1 (T-lymphocyte-specific antigens), treatment with only the divalent polyclonal primary antibody (e.g., mouse anti-histocompatibility or TL or Thy-1) without adding a secondary antibody (anti-antibody, e.g., goat anti-mouse immunoglobulin) induces only patch structures.[12,30] The number of capped cells is greatly enhanced if the cells are treated with both primary and secondary antibodies. At the proper concentration the tetrameric form of concanavalin A (Con A), but not the dimeric form (e.g., succinyl Con A) or the monomeric form, will induce the surface Con A binding sites to aggregate into a cap structure.[32-33] Therefore, the requirement for certain multivalent ligands, possibly serving as a cross linker of the specific receptors, has been strongly implicated in the initial stage of these receptor-ligand redistributions.

Recently, a number of physiologically relevant ligands such as insulin,[34-35] epidermal growth factor,[36] and transferrin[37,38] have been shown to induce their specific receptors to form patches and caps. These receptor clusters and aggregates are very similar to those induced by antibodies and lectins. Since these physiologically relevant ligands are definitely monovalent, the initial receptor aggregation step cannot result from cross linking alone as proposed for antibodies and lectins. Furthermore, *ligand-independent* surface receptor movement has also been reported. For example, treatment of the cells with hypertonic medium, trypsin, cholinergic drugs,[39-41] or colchicine[42] can induce surface receptor patch and cap structures in the absence of any externally added ligands. Therefore, the valency of the ligand may be important in certain cases, but does not appear to be an absolute requirement for the induction of receptor movement in all types of cells.

III. LATERAL MEMBRANE PROTEIN INTERACTIONS

During ligand-induced surface receptor movement there is another important phenomenon called receptor "co-capping". The term co-capping is used to describe lateral membrane protein interactions in which surface proteins are observed to redistribute together. For example, viral antigens are known to co-cap with the H-2 (histocompatability) antigen on the surface of tumor cells[43] and Ig receptor has been found to co-cap with Fc receptors.[44-45] However, Ig also appears to redistribute independently (i.e., does not co-cap) with respect to the C3 receptor, the major histocompatibility antigens (HLA in man and H-2 in mouse) and Ia antigen.[46-49] Furthermore, it has been found that T-200 (gp 180) antigens co-cap with several other antigens such as Thy-1, gp 69/71, and TL antigens on the surface of T-lymphocytes.[30,50]

Recently, insulin receptors have been noted to co-cap with HLA antigens and lectin receptors[35,51] and certain membrane-associated enzymes such as 5'-nucleotidase and ATPase apparently co-cap with Ig receptors.[52] The enzyme, adenylate cyclase, can also be induced to co-cap with Thy-1 antigens.[53] Therefore, it appears that a selective process is involved in the redistribution of some classes of membrane receptors leading to receptor co-capping. Consequently, the biochemical nature of different surface aggregates and their subsequent physiological effects appear to be very dependent on whether the cells are treated with antibody, lectins, or physiologically relevant ligands such as hormones. However, the regulatory mechanisms involved in the redistribution of all surface receptor movement are generally thought to involve an interaction between the surface receptor molecules and underlying cytoskeletal components.

IV. TRANSMEMBRANE INTERACTIONS BETWEEN SURFACE RECEPTORS AND THE CYTOSKELETON

A. Involvement of the Cytoskeleton

1. Microfilaments

Receptor cap structures induced by monovalent ligands (e.g., insulin, EGF, and transferrin), multivalent ligands (e.g., antibodies against specific surface receptors or lectins), or certain nonligand reagents (e.g., hypertonic medium, trypsin, cholinergic drugs, or colchicine) all appear to exhibit the following common characteristics: (1) the capping process is sensitive to microfilament disruption agents, cytochalasin B and D, implying that microfilament function is important for receptor movement;[11,54] (2) patch and cap structures are closely associated with actin, myosin, and actomyosin-associated molecules, but not tubulin;[13,15,55-57] and (3) actin polymerization occurs concomitantly with receptor cap formation.[58] This work provides strong evidence for the direct involvement of microfilaments in the regulation of receptor movement. Detailed information with regard to the molecular interactions between microfilament-associated components and surface molecules will be discussed in Sections IV.B and C.

2. Microtubules

The evidence for the involvement of microtubules in receptor movement, thus far, is rather sketchy and less convincing. In lymphocytes, it is known that low doses of anti-Ig or Con A (below 10 μg/mℓ) induce receptor capping at room temperature of 37°C, but high concentrations of the anti-Ig or lectin (above 50 μg/mℓ) cause the immobilization of their respective receptors under the same incubation conditions.[59] Because this inhibition of receptor capping can be reversed by incubation of the cells in the cold (0 to 4°C) or with microtubule disrupting agents such as colchicine or vinblastine, it has been suggested that microtubules are somehow involved in the modulation of ligand-induced receptor movement.[59] In addition, colchicine not only stimulates ligand-induced capping, but the drug by itself is also able to induce several surface molecules to form cap-like structures in the absence of any external ligand. Colchicine-induced (ligand-independent) capping is cytochalasin D-sensitive.[42] It is possible that normally intact microtubules are always directly or indirectly preventing the contractile activity of microfilaments. After colchicine or vinblastine destroys the integrity of microtubules, strong ligand-receptor interactions appear to somehow overcome the microtubule inhibition of capping.

Recent evidence shows that the 20,000-dalton myosin light chain in lymphocytes is both phosphorylated and preferentially accumulated in the plasma membrane of colchicine-induced capped cells. The phosphorylated myosin light chain may modify the attachment of actomyosin molecules to the plasma membrane which subsequently activates an actomyosin-mediated sliding filament mechanism required for receptor movement. Detailed evidence supporting this possibility is described below. These observations suggest that there is a close interaction between the colchicine-sensitive components (i.e., microtubules) and the actomyosin filaments during receptor movement. The specific molecular interactions between microfilaments and microtubules remain to be elucidated.

3. Intermediate Filaments

Various types of intermediate filaments (approximately 100 Å in diameter) have been observed in a number of different cell types. Based primarily on biochemical and immunological criteria, intermediate filament proteins have been classified into five major groups: vimentin, keratin, desmin, glial filaments, and neurofilaments.[62] In lymphocytes, these intermediate filament proteins are primarily vimentin-like molecules. Morphological analyses indicate that intermediate filaments are often closely associated with various cellular components including the plasma membrane (Figures 1 and 2), the nucleus, and other organelles.[62,63] However, the exact function of intermediate filaments in eukaryotic cells is not known at the present time.

The simultaneous rearrangement of intermediate filaments (e.g., vimentin filaments, keratin filaments) with microtubules and microfilaments has been obseved in canine distemper virus (CDV)-infected and uninfected Hep2 cells using immunofluorescence techniques. The canine distemper virus infection causes a drastic reorganization of all the cytoskeletal structures tested including vimentin and keratin-type intermediate filaments, microfilaments, and microtubules.[64] There is also evidence that cytochalasin D and colchicine are capable of inducing the reorganization of keratin-containing intermediate filaments in cultured epithelial cells.[65] Furthermore, the vimentine-type intermediate filaments appear to interact with microtubules and microfilaments in various cell types.[66-69] Consequently, there seems to be a close association or interaction between intermediate filaments and microfilaments and microtubules.

During ligand-induced receptor capping, intermediate filaments have been shown to accumulate beneath capped structures using both electron microscopic[70] and fluorescence staining techniques.[71-72] Biochemical studies indicate that the vimentin-type of intermediate filaments can be co-isolated with a plasma membrane glycoprotein, gp 140, in cultured human fibroblasts.[73] Therefore, there appears to be a direct physical linkage between intermediate filaments and membrane-associated components.

At the present time, the mechanism determining the location or polarity of the cap structure on the cell surface is unknown. It is interesting that cytoplasts (nucleated cells with only cytoplasm and few intermediate filaments) are unable to form receptor caps whereas the karyoplasts (cells containing a nucleus and perinuclear arrays of intermediate filaments surrounded by a thin rim of cytoplasm) do form normal cap structures. The implication here is that the cell nucleus or some closely associated cellular components may be essential for determining the topographical location of a cap structure on the cell surface. Since intermediate filaments are closely associated with the nucleus[62,63,71] and are concentrated between the surface Con A cap structure and the nucleus,[71] we are suggesting that intermediate filaments may play an important role in directing the polarity of cap formation through their interaction with microfilaments/or microtubules.

B. Structural Linkages Between Surface Receptors and the Cytoskeleton

In all eukaryotic cells that have been studied there appears to be an association between

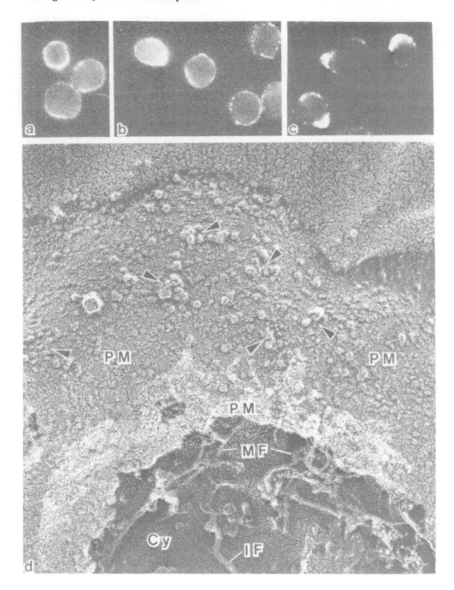

FIGURE 1. Organization of surface molecules visualized by immunofluorescence microscopy (a—c) and freeze-fracture/deep etch technique (d). (a) Prefixed mouse T-lymphoma cells were labeled with rat anti-Thy-1 and fluorescein-conjugated rabbit anti-rat immunoglobulin. Note the uniformly staining pattern of Thy-1 antigens on the cell surface. (b) Unfixed mouse T-lymphoma cells were labeled with both rat anti-Thy-1 plus fluorescein-conjugated rabbit anti-rat immunoglobulin at 0°C for 20 min to induce receptor "patching formation". (c) Unfixed mouse T-lymphoma cells were labeled with both rat anti-Thy-1 plus fluorescein-conjugated rabbit anti-rat immunoglobulin at 37°C for 20 min to induce receptor "cap formation". (d) Anti-Thy-1 antibody treated mouse T-lymphoma cells were quick frozen with a liquid helium cooled machine and freeze-fractured in a Balzers apparatus as described previously.[12] Arrowheads indicate the clustered Thy-1 antigens; PM: plasma membrane; MF: microfilaments; IF: intermediate filaments; Cy: cytoplasm. (Micrographs were obtained as a collaborative effort between Dr. J. Heuser, Washington University, St. Louis, Mo. and our laboratory.)

the surface membrane and the underlying cytoskeleton network (Figures 1d and 2). It has been proposed that a specific transmembrane interaction occurs between intracellular acto-myosin filaments and surface receptors during patching and capping.[12,55] Specifically, this hypothesis states that membrane-associated actin and myosin are either directly or indirectly bound to an integral plasma membrane protein (or class of proteins) called X-protein(s)

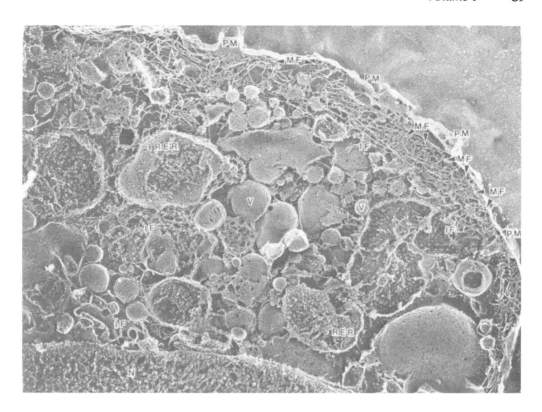

FIGURE 2. Electron micrograph of membrane-cytoskeleton organization in mouse T-lymphoma cells by the freeze-fracture/and deep-etch technique. Anti-Thy-1 antibody treated mouse T-lymphoma cells were deep-etched at −95°C for 5 min. After the samples were cooled to either −120 or −196°C, replicas were rotary shadowed by a mixture of platinum and carbon[12] and examined by a JOEL 100 CX electron microscope. Note the accumulation of microfilaments (MF; arrowheads) beneath the plasma membrane (PM). Many intermediate filaments (IF) were detected in the vicinity of microfilament bundles and among various cellular organelles; vesicles (V), rough endoplasmic reticulum (RER), and nucleus (N). (Micrographs were obtained in collaboration with Dr. J. Heuser, Washington University, St. Louis, Mo.)

following the binding of an external ligand (e.g., antibodies against specific receptors or lectins) to the cells and during the subsequent aggregation of the surface receptors into "patches". These receptor aggregates (linked to actin and myosin through X-protein(s)) are then collected into a cap by a sliding filament mechanism analogous to that occurring during muscle contraction.[12,55] In recent years some progress has been made in the identification of the putative X protein(s) and its linkage with the cytoskeleton as well as the analysis of the muscle-like contractility occurring during receptor movement.

Currently, the most well-defined membrane-cytoskeleton organization is that which occurs in erythrocytes and intestinal brush border cells. In erythrocytes, the linkage between band 3 membrane protein (an anion transport channel protein) and spectrin (a cytoplasmic actin binding protein) is mediated by the membrane attachment protein, ankyrin.[78,79] The association between glycophorin (a major erythrocyte membrane glycoprotein) and actin is via band 4.1 protein.[80] In brush border cells recent evidence indicates that the 110,000-dalton actin-binding protein is also an integral membrane protein.[81]

Recently, analogs of the erythrocyte membrane-cytoskeleton proteins, such as spectrin, band 4.1 protein, and ankyrin, have been found in nonerythroid cells. Two distinct spectrin-like proteins have been identified: fodrin,[82,83] isolated from brain and TW 260/240,[84] isolated from epithelial cells of the intestinal brush border. Each of these proteins has one 240,000-dalton subunit and a second subunit of differing molecular weights: 220,000 daltons for

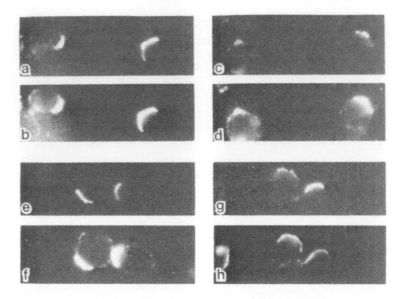

FIGURE 3.   Simultaneous localization of receptor caps and intracellular cytoskeletal proteins. Lymphocyte surface receptors were first labeled with fluorescein-wheat germ agglutinin (Fl-WGA) in the capped condition (a, c, e, and g) followed by intracellular labeling of rhodamine (Rh)-conjugated anti-actin (b), Rh-conjugated-anti-fodrin (d), rhodamine-conjugated anti-band 4.1 (f), and rhodamine-conjugated anti-ankyrin (h). In each pair of figures (a and b); (c and d); (e and f); and (g and h) the Fl-WGA caps in panel a, c, e, and g are from the same cells as shown in the corresponding panel b, d, f, and h, respectively.

spectrin,[85] 235,000 daltons for fodrin,[82] and 260,000 daltons for TW 260/240.[83] Immunological analysis and protein peptide mapping data indicate that the 240,000-dalton subunits of spectrin, fodrin, and TW 260/240 are very similar.[82,83]

Band 4.1 protein (M_r 80,000 daltons) in erythrocytes enhances the binding of spectrin to actin[86-88] and also binds to transmembrane proteins.[89-91] Immunoreactive forms of band 4.1 protein appear during SDS-gel electrophoresis as a doublet with apparent molecular weights ranging from 68,000 to 75,000 daltons in brain[92] and other tissues.[93] Brain 4.1-like protein is identical in M_r on SDS analysis, physical properties, immunoreactivity, and peptide maps to synapsin, one of the major components of synaptic vesicles.[94]

Ankyrin, one of the linker molecules bridging spectrin and membrane proteins has been shown to be a single polypeptide of M_r 215,000 daltons in erythrocytes[95] and two polypeptides of M_r 220,000 daltons and 210,000 daltons in brain.[96] Restricted proteolytic digestion of isolated ankyrin has been used to identify at least two functional domains of the molecule. The 72-kdalton chymotryptic or 65/55-kdalton tryptic fragments bind spectrin whereas the 82/83-kdalton chymotryptic and tryptic fragments bind to the cytoplasmic domain of band 3 protein.[97-99]

Lymphocyte surface membrane is attached to actomyosin (Figure 3b) as well as immunoreactive forms of fodrin (Figure 3d), band 4.1 protein (Figure 3f), and ankyrin molecules (Figure 3h), all of which co-cap with surface receptor cap-structure (Figure 3a, c, e, and g). To biochemically determine whether any particular plasma membrane proteins are tightly associated with the lymphocyte cytoskeleton, cells were first metabolically labeled with ^{3}H-glucosamine. SDS-gel electrophoresis and autoradiographic analyses of the total lymphocyte plasma membrane preparation reveal many ^{3}H-glucosamine-labeled polypeptides with molecular weights from 250,000 to 25,000 daltons plus several major glycoproteins — gp 180, gp 85, and gp 25 (Figure 5 gel lane). Sucrose gradient centrifugation of the nonionic detergent

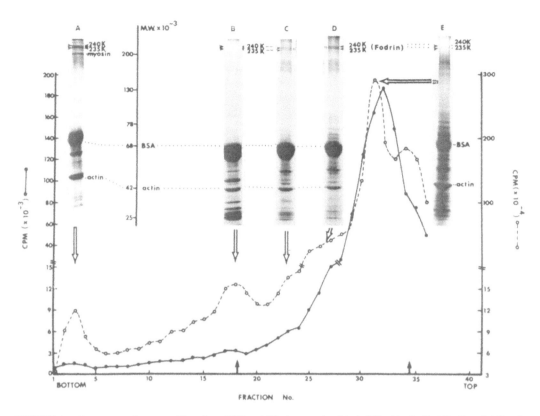

FIGURE 4.   Sucrose gradient centrifugation of ^{125}I and ^{3}H-glucosamine-labeled Nonidet P-40 (NP-40) solubilized mouse T-lymphoma plasma membrane. The distribution of plasma membrane proteins labeled with [125]I (○– – – –○, representing total proteins) or ^{3}H-glucosamine (●———●, indicating associated glycoproteins) within the 8 to 47% sucrose gradient (centrifuged at 70,000 × G_{av} for 12 hr) is plotted as counts per minute vs. fraction number. The fractions that constitute individual peaks of protein were pooled, the proteins precipitated with 10% trichloroacetic acid in the presence of bovine serum albumin (BSA), and then analyzed by SDS-PAGE. Peak A = fractions 1 to 5; peak B = fractions 16 to 20; peak C = fractions 25 to 27; peak D = fractions 28 to 29; peak E = fractions 30 to 35. The arrows at fraction numbers 18 and 34 indicate the respective positions of native horse ferritin (2,500,000 mol wt) and pig brain fodrin (950,000 mol wt) as molecular weight markers in the gradient. (The fact that actin was detected in all peaks suggested the possible existence of various sizes of actin-containing filaments. Fodrin and other unidentified actin-binding proteins were selectively associated with certain sizes of actin-containing filaments (e.g., in peaks A, C, D, E but not others e.g., not in peak B).

(NP-40 or Triton X-100) solubilized lymphocyte plasma membrane fraction indicates that there is a close association of some glycoproteins with various sizes of actin-containing filaments (Figure 4). GP 25 is selectively distributed with the larger-sized actin-containing filaments (Figure 5), whereas gp 180 and gp 85 appear to be preferentially associated with the smaller-sized actin-containing filaments (Figure 5). These data suggest that there is a differential distribution of various glycoproteins with different sizes of actin-containing filaments. Further analysis shows that lymphocyte gp 180, a transmembrane glycoprotein, is preferentially associated with the actin-binding protein, fodrin, in a 1:1 molar ratio. This complex, displaying a sedimentation coefficient of approximately 20 S, remains stable during isoelectric focusing and exhibits a pI in the range of 5.2 to 5.7.[100] The fact that this gp 180-fodrin complex can be dissociated and reassociated in the presence and absence of high salt[101] indicates that the binding of these two molecules is rather specific.

Most recently, we have found that another lymphocyte transmembrane glycoprotein, gp 85, is tightly associated with an ankyrin-like protein in a 16 S complex that resembles the erythrocyte band 3-ankyrin complex.[102] This gp 85-ankyrin complex does not separate in

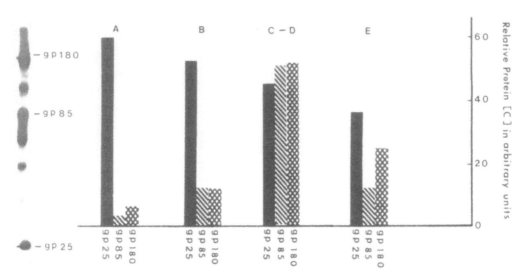

FIGURE 5.   Quantitation of the relative amounts of specific glycoproteins in peaks A, B, C-D, and E obtained from a sucrose gradient similar to that described in Figure 4. The gel lane illustrates the major ^{3}H-glucosamine-labeled NP-40 solubilized plasma membrane glycoproteins, gp 180, gp 85, and gp 25, using SDS-PAGE and autoradiographic analysis. The values obtained from each peak are normalized per milliliter fraction and are expressed as relative protein concentration. (Note the differential distribution of gp 180, gp 85, and gp 25 in the various actin-containing filament fractions suggesting a differential association of glycoproteins with varying sizes of actin filaments)· gp 180 = ▓▓▓, gp 85 = ▨▨, and gp 25 = ■ .

the presence of high salt but can be dissociated by treatment with 2 *M* urea. The interaction between these two transmembrane complexes during lymphocyte membrane activity is not clear at this time. Although an ankyrin-like protein has not been identified in the gp 180-fodrin complex, it is possible that gp 180 either contains an ankyrin-like domain or interacts with a small fraction of the ankyrin-gp 85 complex or ankyrin alone which then is responsible for the subsequent linkage between gp 180 to fodrin. Most importantly, during ligand-induced capping there is a parallel recruitment of both gp 180-fodrin and gp 85-ankyrin complexes into the actin-containing cytoskeleton fraction and the amount of these two transmembrane complexes is directly proportional to the percentage of capped cells.[100,102] Therefore, the involvement of transmembrane complexes in receptor movement is strongly implicated. An independent study has determined that a major platelet surface glycoprotein, gp I, is also closely associated with the actin-binding protein, filamin.[103] All of this evidence supports the notion that certain linker molecules (e.g., ankyrin) and actin binding proteins (e.g., fodrin, 4.1-like proteins and filamin) play a pivital role in linking surface receptors to the intracellular actin-containing microfilament network.[12]

Recently, it has also been found that the tumor-promoting agent, phorbol-12-*O*-tetra-decanoylphorbol 13 acetate (TPA), is able to activate protein kinase C which, in turn, phosphorylates various proteins such as myosin light chain,[104] tropomyosin,[105] and the trans-membrane glycoprotein, gp 180.[106] The phosphorylated myosin light chain, tropo-myosin, and gp 180 all appear to be preferentially linked to the cytoskeleton as well as being involved in platelet activation[104-105] and lymphocyte capping.[107] Consequently, we believe that protein kinase C plays an important role in the selective activation of various cellular constituents responsible for the physiological responses associated with ligand-receptor in-teractions. It is possible that, after ligand binding to surface receptors, the receptors are first modified by phosphorylation and then bind effectively to an actin-binding protein complex. Due to the cross linking of certain actin-binding protein(s) to actin filaments and the sub-sequent contraction of actomyosin (analogous to the sliding filament mechanism during smooth muscle contraction), the aggregates (patches) of ligand-receptor complexes are collected into a cap structure.[12,55]

C. Regulation of Receptor Movement By a Mechanism Analogous To Muscle Contraction

1. The Requirement for Metabolic Energy (ATP)

Cellular ATP production is known to be controlled by two major metabolic pathways, oxidative phosphorylation and glycolysis. The commonly used oxidative phosphorylation inhibitors such as sodium azide, potassium cyanide, oligomycin, and dinitrophenol have been shown to effectively inhibit the formation of cap structures (but not patch structures). In contrast, glycolysis inhibitors such as 2-deoxyglucose, fluoride, and iodacetamide appear to have very little effect on the receptor movement process.[108] These data suggest that receptor movement requires energy derived primarily from oxidative phosphorylation, but not from glycolysis.

In order to demonstrate the direct involvement of ATP in ligand-induced capping, we have developed a procedure that involves the use of an EGTA-buffered solution to render the cell membrane reversibly permeable to ATP.[109] Our results show that receptor capping can be induced in the presence of exogenous ATP and a high Ca^{2+} concentration (10^{-4} M), whereas in the absence of exogenous ATP (but with high Ca^{2+} concentration), cap formation is inhibited.[109] In muscle cells the energy requirement for the cyclic attachment of myosin cross bridges to actin filaments is provided by ATP.[110] Removal of ATP, in the presence or absence of Ca^{2+}, results in a noncyclic interaction of actin and myosin which prevents further contraction or relaxation of muscle (rigor state). This rigor state can only be overcome by the re-addition of ATP. Like muscle relaxation, cap structure dissociation is prevented in the absence of ATP and occurs in the presence of ATP.[109] In the presence of ATP, analogs such as UTP, ITP, CTP, GTP, GTP [γ S] (5'-[γ-thio]triphosphate) and p[NH]ppA (5'-adenylyl imidodiphosphate) surface receptor movement does not occur.[109] The specific ATP requirement for receptor cap formation and dissociation is consistent with the hypothesis that receptor movement is regulated by a mechanism analogous to the sliding-filament mechanism for muscle contraction and relaxation.

2. The Regulatory Roles of Ca^{2+}, Calmodulin, and Myosin Light Chain Kinase

a. Calcium (Ca^{2+})

Cytosolic free calcium, either entering through the plasma membrane or released from intracellular stores, has been proposed to be a regulator of a large number of cellular activities.[111] External ligands such as hormones and neurotransmitters, that use Ca^{2+} as a second messenger, cause the hydrolysis of membrane phosphoinositides into diacylglycerol and inositol triphosphate (IP_3). Diacylglycerol has been suggested to stimulate protein phosphorylation, whereas IP_3 has been implicated in the release of Ca^{2+} from intracellular storage sites.[112] While there is ample evidence for a rise in intracellular Ca^{2+} concentration following ligand binding to cell surface receptors, it is still unclear whether an influx of extracellular Ca^{2+} or the release of Ca^{2+} from intracellular storage compartments is responsible for the observed cellular activation. Nevertheless, there is a strong correlation between increased Ca^{2+} activity and ligand-induced receptor movement.[35,109,113-115]

Previously, it had been very difficult to quantitatively monitor changes in intracellular calcium concentrations following ligand binding. However, Tsien and co-workers have recently developed a new, calcium-sensitive fluorescent indicator, quin 2 [quin 2-tetra (acetoxymethyl) ester], which responds to changes in the free calcium concentration inside intact cells.[114] Using quin 2, Pozzan et al.[115] reported that there is a rise in internal free calcium from approximately 120 to 500 nM within minutes following the addition of immunoglobulin to mouse splenic lymphocytes. Quin 2 has also been used effectively to monitor intracellular free calcium concentrations in platelets[116-117] following the binding of external ligands such

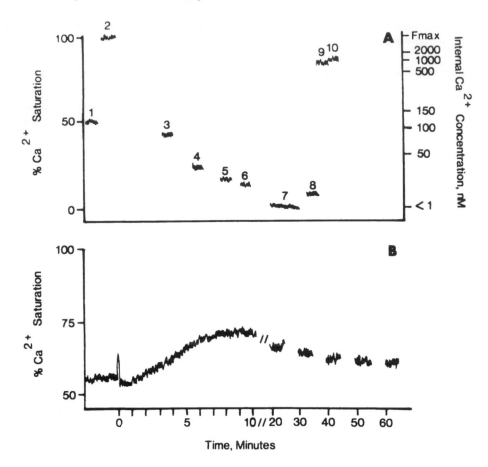

FIGURE 6.   Fluorometric analysis of quin 2 fluorescence. (A) Calibration of quin 2-loaded human B-lymphoblasts was carried out as follows. Characterization of a typical batch of cells is shown: (1) quin 2-loaded intact cells; (2) maximum fluorescence after digitonin permeabilization in the presence of 1 mM calcium; (3 to 6) subsequent additions of 2 mM EGTA to lower the free calcium concentration with corresponding decreases of fluorescence. The final EGTA concentrations are (3) 2 mM, (4) 4 mM, (5) 6 mM, and (6) 8 mM, and (7) minimal fluorescence when the calcium concentrations is < 1 nM in the presence of Tris base at pH 8.3. As calcium is added, bringing its concentration to (8) 7 mM, (9) 9 mM, and 11 mM, the fluorescence approaches its maximum again. The scale of calcium saturation corresponds to changes between 100% saturation = F_{max} and 0% saturation = F_{min}, as read directly from the fluorometer. The internal calcium concentration is determined from the equation: $Ca^{2+}_i = K_D (F-F_{min})/(F_{max}-F)$ with K_D reported at 115 nM.[113-114] (B) Goat anti-human immunoglobulin (50 to 100 μg/mℓ) was added to a suspension of cells loaded with quin 2 as described above. By 1 min, the fluorescence began to increase indicating a rise in free internal calcium concentration which plateaus by 8 min and then beginning a slow descent.

as thrombin[116] and platelet activating factor.[118] In addition, there is preferential localization of intracellular Ca^{2+} in human neutrophils during chemotaxis, oxidative activity, and degranulation as measured by quin 2.[121] In our studies we have found that the characteristics of quin 2 loading and calibration with human B lymphoblasts are similar to those cell types previously reported (Figure 6A). In addition, we have found that during Ig capping the quin 2 fluorescence signal begins to increase within 1 min of ligand addition, rises to over four times its baseline level by 8 min, and then begins a slow decrease (Figure 6B). Using fluorescence microscopy we were then able to distinguish a preferential accumulation of intracellular Ca^{2+} (indicated by a localized quin 2 fluorescence) underneath anti-Ig-induced receptor cap structures (Figure 7b) as compared to uniformly distributed Ca^{2+} (indicated by

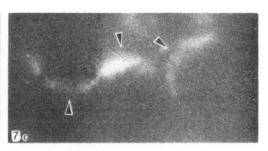

FIGURE 7. Fluorescence microscopy of quin 2-loaded human B lymphoblasts. Under fluorescent microscopic examination, quin 2-loaded cells showed a diffuse fluorescence with the cytoplasmic side slightly brighter (a) When stimulated by goat anti-human immunoglobulin (b) or insulin (c), however, the fluorescent signal became more concentrated at one side of the cell apparently under the plasma membrane (▼). (Using a double immunofluorescence staining, we observed the quin 2 fluorescence concentrated directly beneath receptor capped structures (not shown).

diffuse quin 2 fluorescence) in untreated cells (Figure 7a). A similar localization of internal Ca^{2+} can also be observed upon stimulation of human B lymphoblasts with insulin (Figure 7c). Therefore, an important role for Ca^{2+} in the mechanism of cell activation following ligand-receptor binding is strongly implicated.

To directly test the role of Ca^{2+} in receptor capping we have utilized the previously mentioned EGTA-buffered permeabilizing solution to render the membranes reversibly permeable to exogenously added Ca^{2+}. We then monitored the effect of varying Ca^{2+} concentrations on the degree and extent of capping.[109] Our results show that there is an increase in the number of capped cells proportional to the increase in Ca^{2+} concentration between 10^{-6} and $10^{-4.9}$ M. It is also interesting that the threshold Ca^{2+} concentration (10^{-6} M) required for the activation of the capping process is almost identical to that required for the activation of contraction in smooth and striated muscle.[110] Clearly, there are some striking similarities between the requirements for both muscle contraction and receptor movement.

Related findings have been reported by Haslam and Davison using an electric discharge method to permeabilize blood platelets.[122] Their results indicate that the addition of micromolar concentration of Ca^{2+} causes an ATP-dependent secretion of serotonin from the platelet-dense granules.[122] Therefore, it is suggested that Ca^{2+} is one of the primary triggers for cell activation.

b. Calmodulin

Calmodulin is one of several Ca^{2+} binding proteins that is known to be involved in cellular Ca^{2+} regulation.[123] This protein contains about a 70% amino acid sequence homology with troponin C, a regulatory protein involved in the actomyosin contractile mechanism.[123] In order to directly demonstrate the participation of calmodulin during receptor-ligand interaction, we again made use of the EGTA-buffered permeabilizing solution.[109,124] Our results demonstrate that the addition of calmodulin (at the minimal Ca^{2+} concentration [10^{-6} M] required to activate receptor movement) results in a significant stimulation of capping.[109] Furthermore, an increase in the calmodulin concentration from 3.9×10^{-8} to 6.3×10^{-7} M causes a proportional increase in the number of capped cells.[124] In the absence of Ca^{2+}, calmodulin has no effect on receptor movement[109,124] Several antipsychotic drugs, including chloropromazine (thorazine) and trifluoroperazine (stelazine) as well as the W series of drugs (W 5, W 7, and W 12), are known to impair the function of calmodulin.[125-127] These inhibitors have also been shown to block ligand-induced receptor capping[126-127] and smooth muscle contraction[128] in a dose-dependent manner in either intact or EGTA-buffer permeabilized cells. Using a double immunofluorescence technique, both intracellular Ca^{2+} and calmodulin were found to accumulate underneath receptor cap structures.[51,127] All of the aforementioned evidence supports the contention that a Ca^{2+}/calmodulin complex is responsible for the activation of receptor movement analogous to smooth muscle contraction.

c. Myosin Light Chain Kinase (MLCK)

Studies on both smooth muscle and nonmuscle cells indicate that actomyosin-mediated contraction is regulated by myosin light chain kinase (MLCK) which in turn is regulated by Ca^{2+}/calmodulin.[129-130] In addition, the phosphorylation of myosin light chain has been shown to enhance the actin-activated Mg^{2+}-ATPase of myosin which produces the energy to drive the sliding filament contraction process.[131-132] In what might be an analogous process, the 20,000-dalton myosin light chain in lymphocytes becomes phosphorylated following receptor-ligand binding.[42,133] Additionally, it has been shown that receptor movement *in vitro* can be directly regulated by micromolar concentrations of Ca^{2+} and calmodulin.[109,124] These results, coupled with the myosin light chain phosphorylation data, strongly support the contention that this nonmuscle contractile process (receptor patching/capping) is MLCK-dependent. In blood platelets, it has been reported that the release of serotonin from permeabilized or intact platelets is also associated with the Ca^{2+}-dependent phosphorylation of the 20,000-dalton myosin light chain-like protein.[104-106,122]

A specific antibody raised against chick gizzard MLCK[133] has been utilized to identify the presence of a MLCK in mouse T-lymphoma cells. The lymphocyte MLCK is found to be a 130,000-dalton protein that is present in plasma membrane-cytoskeleton preparations. During ligand-induced receptor capping, the MLCK is also preferentially accumulated under the surface cap-structure. Furthermore, the lymphocyte MLCK can phosphorylate both endogenous lymphocyte myosin light chain and myosin light chains from smooth and skeletal muscle in a Ca^{2+}/calmodulin-dependent and trifluoperazine-sensitive manner.[133]

During muscle contraction MLCK is known to utilize both ATP and adenosine 5'-[γ-thio]triphosphate (ATP [γ S]) (at high Ca^{2+} concentration [$10^{-4}M$]) as substrates to phosphorylate or thiophosphorylate myosin light chains. Normally, phosphorylated myosin light chains are readily dephosphorylated by phosphatase enzyme resulting in muscle relaxation.[129] In contrast, thiophosphorylated myosin light chains are resistant to the phosphatase resulting in an irreversible contraction state of rigor.[128] To test whether a similar MLCK-mediated activity is regulating receptor movement in nonmuscle cells, lymphocytes were incubated in the EGTA-buffered permeabilizing solution containing active MLCK, various Ca^{2+} concentrations (pCa = 4.0 or pCa = 8.0), and either ATP or ATP [γ S]. Our data show that the formation of receptor caps was signficantly stimulated in the presence of a high concentration of Ca^{2+} (pCa = 4.0) with ATP or ATP [γ S].[124] However, the dissociation of cap structures was observed only in the presence of ATP but not ATP [γ S]. These results imply that the cyclic events of phosphorylation/dephosphorylation of myosin light chains are probably responsible for cap formation and cap dissociation.[124] The failure of phosphatase to dephosphorylate the thiophosphorylated myosin light chains would explain why the cap structures failed to be dissociated.[124] This irreversible capping phenomenon is exactly analogous to that occurring during irreversible contraction in muscle cells.[128]

In smooth muscle there is also a transient increase in intracellular Ca^{2+} level just prior to contraction.[134] A similar pattern, namely a transient increase in Ca^{2+} level preceding receptor capping, has been observed during receptor-ligand interaction.[115] Biochemical studies using SDS gel electrophoresis and autoradiography reveal that the myosin light chain is preferentially phosphorylated in the presence of $AT^{32}P$ and high concentrations of Ca^{2+} (pCa = 4.0) and calmodulin (Figure 8b), but not in the presence of trifluoroperazine (a calmodulin inhibitor) (Figure 8c) or in the absence of Ca^{2+} (pCa = 8.0) (Figure 8a). Therefore, a Ca^{2+}/calmodulin-dependent MLCK-mediated myosin light chain phosphorylation appears to be definitely required for receptor movement. Clearly, several characteristics of receptor capping

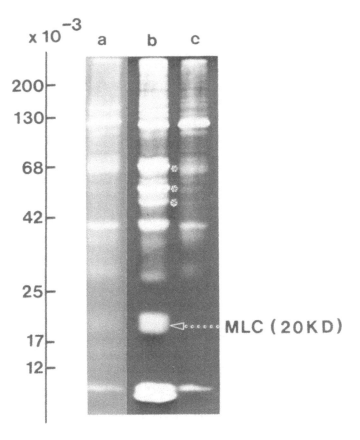

FIGURE 8.   Detection of myosin light chain (MLC) phosphorylation. Fluorescent concanavalin A (Fl-Con A) labeled mouse T-lymphoma cells were incubated in an EGTA-buffered permeablizing solution[124] containing AT^{32}P in the presence or absence of Ca^{2+}/calmodulin and calmodulin inhibitor, trifluoperazine (TFP, stelazine) for 10 min at room temperature. An aliquot of cells was then removed and fixed in 2% paraformaldehyde to monitor the degree of capping. The remaining ^{32}P-labeled cells were extracted by nonionic detergent (1% NP-40) followed by 100,000 × g$_{av}$ centrifugation at 4°C for 30 min. The pelletable material (detergent-insoluble cytoskeleton)[100,133] was then analyzed by SDS-PAGE and autoradiographic analyses. (a) When cells were incubated in an EGTA-buffered permeabilizing solution containing pCa = 8.0 and AT^{32}P, few Con A capped structures were observed. Note a number of phosphoproteins were detected. However, there was an absence of phosphorylated proteins in the 20-kdalton polypeptide region. (b) When cells were incubated in an EGTA-buffered solution containing pCa = 4.0, calmodulin, and AT^{32}P, a maximal number of Con A capped structures were observed. (Note that several proteins as indicated by asterisks and a 20-kdalton polypeptide were preferentially phosphorylatted (▲). The 20-kdalton phosphoproteins were identified as an MLC by an anti-myosin antibody; data not shown). (c) When cells were incubated in an EGTA-buffered permeablizing buffer containing pCa = 4.0, calmodulin, and trifluoperazine, Con A cap formation was inhibited. (Note the inhibition of phosphorylation in the 20-kdalton polypeptide region.)

lymphocyte membranes bear striking similarities to the mechanisms underlying contraction in smooth muscle. A conceptual model which has been proposed to explain activation of smooth muscle contraction has been extended to receptor capping as shown in Figure 9.

V. SCHEMATIC MODEL OF LIGAND-INDUCED SURFACE RECEPTOR MOVEMENT

Based on the information presented above, our current ideas concerning the sequence of

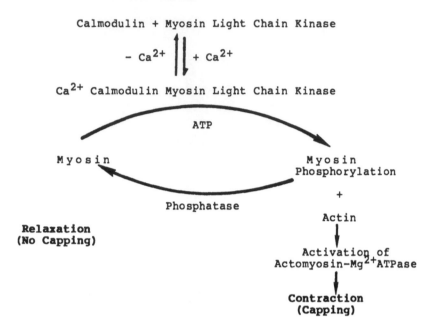

FIGURE 9.   Model for the activation of smooth muscle contraction and receptor capping.

events involved in ligand-induced surface receptor movement are summarized schematically in Figure 10.

ACKNOWLEDGMENTS

The author is grateful to Drs. Gerard J. Bourguignon and Suzanne J. Suchard for their helpful discussion and assistance in preparing this manuscript. In addition, I would like to thank Dr. Suzanne J. Suchard and Dr. Mary Majercik for providing Figures 4 and 5 and Figure 6 and 7, respectively. This work was supported by grants from U.S. Public Health Service GM 36353 and AI 19188 and the American Heart Association. L. Y. W. B. is an Established Investigator from the American Heart Association.

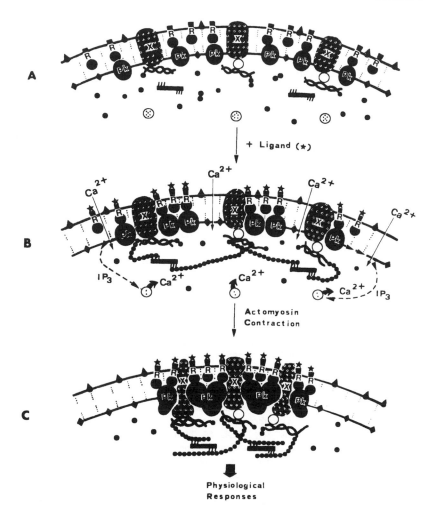

FIGURE 10.   Schematic model of ligand-induced surface receptor movement. (A) Membrane and cytoskeletal components in the ligand-free (resting) state. In the ligand-free resting state membrane components such as surface receptors (R) (e.g. surface antigens and hormone receptors), X protein(s) (X), protein kinase(s) (pK) (e.g. protein kinase C, tyrosin kinase, and cAMP-dependent kinase) and other membrane proteins (♠ , ♦) are uniformly distributed in the plane of the plasma membrane. Important cytoskeletal proteins include actin (●, primarily in G-form), myosin ( ), fodrin ( ), and membrane attachment proteins (○, such as ankyrin-like and/or 4.1-like proteins). Note that fodrin and X protein(s) are associated either through direct binding or membrane attachment proteins. ⊛   , indicates cytoplasmic Ca^{2+} storage compartments such as smooth endoplasmic reticulum-like vesicles. (B) Ligand-induced surface receptor patching and transmembrane signaling. Upon binding of ligand (★) (e.g. multivalent lectins, antibodies or monovalent hormones, growth factors) to specific receptors (R), there is initially spontaneous patch formation of certain membrane components including the ligand-bound receptors (R), X protein(s) (X), and adjacent protein kinase(s) (pK). G-actin is induced to polymerize into F-actin which becomes associated with myosin ( ). F-actin also binds to X protein(s) via fodrin ( ) or fodrin-associated membrane attachment proteins (○, ankyrin-like or 4.1-like protein). These events are accompanied by a rise in the intracellular free Ca^{2+} level possibly due to an increased Ca^{2+} influx or an IP₃ (inositol-1,4,5-triphosphate)-mediated or possibly some cytoplasmic factor-regulated Ca^{2+} release from cytoplasmic storage compartments ( ). (C) Surface receptor movement into a cap structure. Incubation of ligand-bound (R) cells at room temperature of 37°C activates the Ca^{2+}/calmodulin-dependent MLCK (Figure 9). MLCK-mediated phosphorylation of myosin light chain allows the actin-activated Mg^{2+}-ATPase of myosin to release the energy required to drive the sliding filament contraction process ( ). This actomyosin-mediated contraction then brings all the patched membrane components into one large aggregate (cap formation). It is speculated that this aggregation of membrane protein capping induces changes in the conformation and activity of the associated protein kinase(s) (pk) (e.g., protein kinase C, tyrosin kinase, and cAMP-dependent kinase etc.), which then leads to subsequent physiological responses such as cell activation, mitogenesis, and secretion.

REFERENCES

1. **Heidman, T. and Changeux, J. P.**, Structure and functional properties of the acetylcholine receptor protein in its purified and membrane bound states, *Annu. Rev. Biochem.*, 47, 317, 1978.
2. **DeMeyts, P.**, Cooperative properties of hormone receptors, *J. Supramol. Struct.*, 4, 241, 1976.
3. **DeLisi, C.**, The biophysics of ligand-receptor interactions, *Q. Rev. Biophys.*, 13, 201, 1980.
4. **Fanger, M. W., Hart, D. A., Wells, J. V., and Nisonoff, A.**, Requirement for cross-linkage in the stimulation of transformation of rabbit peripheral lymphocytes by antiglobulin reagents, *J. Immunol.*, 105, 1484, 1970.
5. **Krakauer, H., Peacock, J. S., Archer, B. G., and Krakauer, T.**, The interaction of surface immunoglobulins of lymphocytes with highly defined synthetic antigens, in *Physical Chemical Aspects of Cell Surface Events in Cellular Regulation*, Elsevier/North-Holland, Amsterdam, 1979, 345.
6. **Ishizaka, K. and Ishizaka, T.**, Induction of erythema-wheal reaction by soluble antigen-antibody complexes in humans, *J. Immunol.*, 101, 68, 1968.
7. **Siraganian, R. P., Hoak, W. A., and Levine, R. B.**, Specific *in vitro* histamine release from basophils by bivalent haptens: evidence for activation by simple bridging of membrane bound antibody, *Immunochemistry*, 12, 149, 1975.
8. **Kahn, C. R., Baird, K. L., Jarrett, D. B., and Flier, J. S.**, Direct demonstration that receptor cross-linking or aggregation is important in insulin action, *Proc. Natl. Acad. Sci. U.S.A.*, 75, 4209, 1976.
9. **Schechter, Y., Hernaez, L., Schlessinger, J., and Cautrcasas, P.**, Local aggregation of hormone receptor complexes is required for activation of epidermal growth factor, *Nature (London)*, 278, 835, 1979.
10. **Sedlacek, H. H., Stark, J., Seiler, F. R., Ziegler, W., and Wiegandt, H.**, Cholera toxin induced redistribution of sialoglycolipid receptor at the lymphocyte membrane, *FEBS Lett.*, 61, 272, 1976.
11. **Taylor, R. B., Duffus, P. H., Raff, M. C., and de Petris, S.**, Redistribution and pinocytosis of lymphocyte surface immunoglobulin molecule induced by anti-immunoglobulin antibody, *Nature (London) New Biol.*, 233, 225, 1971.
12. **Bourguignon, L. Y. W. and Bourguignon, G. J.**, Capping and the cytoskeleton, *Int. Rev. Cytol.*, 87, 195, 1984.
13. **Bourguignon, L. Y. W.**, Receptor capping in platelet membranes, *Cell Biol. Int. Rep.*, 8, 19, 1984.
14. **Edidin, M. and Weiss, A.**, Antigen cap formation in cultured fibroblasts: a reflection of membrane fluidity and of cell motility, *Proc. Natl. Acad. Sci. U.S.A.*, 69, 2456, 1972.
15. **Bourguignon, L. Y. W. and Rozek, R. J.**, Capping of concanavalin A receptors and its association with microfilaments in monolayer grown human fibroblastoid cells, *Cell Tissue Res.*, 205, 77, 1980.
16. **Edelman, G. M. and Yahara, I.**, Temperature-sensitive changes in surface modulating assemblies of fibroblasts transformed by mutants of Rous sarcoma virus, *Proc. Natl. Acad. Sci. U.S.A.*, 73, 2049, 1976.
17. **Sasaki, J., Kanda, S., Otsuka, N., Nakamoto, S., and Mori, M.**, Morphology of cap formation in Ehrlich ascites tumor cells induced by concanavalin A, *Cell Struct. Function*, 4, 1, 1979.
18. **Oppenheimer, S. B., Bales, B. L., Brenneman, G., and Knapp, L., Lesin, E. S., Neri, A., and Pollock, E. G.**, Modulation of agglutinability by alteration of the surface topography in mouse ascites tumor cells, *Exp. Cell Res.*, 105, 291, 1977.
19. **Comoglio, P. M. and Filogamo, G.**, Plasma membrane fluidity and surface motility of mouse C-1300 neuroblastoma cells, *J. Cell Sci.*, 13, 415, 1973.
20. **Hooghe-Peters, E. L. and Hooghe, R.**, Capping of nerve cell antigens, *Immunol. Lett.*, 2, 71, 1980.
21. **Albertini, D. F. and Anderson, E.**, Microtubule and microfilament rearrangements during capping of concanavalin A receptors on cultured ovarian granulosa cells., *J. Cell Biol.*, 73, 111, 1977.
22. **Condeelis, J.**, Isolation of concanavalin A caps during various stages of formation and their association with actin and myosin., *J. Cell Biol.*, 80, 751, 1979.
23. **Besterman, J. M. and Low, R. B.**, Endocytosis: a review of mechanisms and plasma membrane dynamics, *Biochem. J.*, 210, 1, 1983.
24. **Ben-Bassat, H. and Goldblum, N.**, Con A receptors on the surface membrane of lymphocytes from patients with Hodgkin's disease and other malignant lymphomas, *Proc. Natl. Acad. Sci. U.S.A.*, 72, 1046, 1975.
25. **Cohen, H. J.**, Human lymphocyte surface immunoglobulin capping, *J. Clin. Invest.*, 55, 84, 1975.
26. **Kammer, G. M.**, Impaired T-cell capping and receptor regeneration in active systemic lupus erythematosus, *J. Clin. Invest.*, 72, 1686, 1983.
27. **Naeim, F., Bergmann, K., and Gatti, R. A.**, Membrane receptors and their redistribution in lymphoproliferative disorders, *Blood*, 54, 648, 1979.
28. **Pickard, N. A., Gruemer, H. D., Verrill, H. L., Isaacs, E. R., Robinow, M., Nance, W. E., Myers, E. C., and Golsmith, B.**, Systematic membrane defect in the proximal muscular dystrophies, *New Engl. J. Med.*, 299, 841, 1978.
29. **Quagliata, F. and Karpatikin, S.**, Impaired lymphocyte transformation and capping in autoimmune thrombocytopenic purpura, *Blood*, 53, 341, 1979.

30. **Bourguignon, L. Y. W., Hyman, R., Trowbridge, I., and Singer, S. J.,** The participation of histocompatibility antigens in the capping of molecularly independent cell surface components by their specific antibodies, *Proc. Natl. Acad. Sci. U.S.A.,* 75, 2406, 1978.
31. **Schreiner, G. F. and Unanue, E. R.,** Membrane and cytoplasmic changes in lymphocytes induced by ligand-surface immunoglobulin interaction, *Adv. Immunol.,* 4, 37, 1976.
32. **Edelman, G. M., Yahara, I., and Wang, J. L.,** Receptor mobility and receptor-cytoplasmic interactions in lymphocytes, *Proc. Natl. Acad. Sci. U.S.A.,* 70, 1442, 1973.
33. **Gunther, G. R., Wang, J. L., Yahara, I., Cunningham, B. A., and Edelman, G. M.,** Concanavalin A derivatives with altered biological activities, *Proc. Natl. Acad. Sci. U.S.A.,* 70, 1012, 1973.
34. **Schlessinger, J., Van Obberghen, E., and Kahn, C. R.,** Insulin and antibodies against insulin receptor cap on the membrane of cultured human lymphocytes, *Nature (London),* 286, 729, 1980.
35. **Majercik, M. H. and Bourguignon, L. Y. W.,** Insulin receptor capping of its correlation with calmodulin-dependent myosin light chain kinase, *J. Cell. Physiol.,* 124, 403, 1985.
36. **Schreiner, A. B., Libermann, T. A., Lax, I., Yarden, Y., and Schlessinger, J.,** Biological role of epidermal growth factor-receptor clustering, *J. Biol. Chem.,* 258, 846, 1983.
37. **Enns, C. A., Larrick, J. W., Suomalainen, H., Schroder, J., and Sussman, H. H.,** Co-migration and internalization of transferrin and its receptor on K 562 cells, *J. Cell Biol.,* 97, 579, 1983.
38. **Bourguignon, L. Y. W., and Morejon, O.,** Transferrin mediated endocytosis in lymphoid cells, in *Third Congress on Cell Biology,* Academic Press, New York, 1984, 344.
39. **Yahara, I. and Kakimoto-Sameshima, F.,** Ligand-independent cap formation: redistribution of surface receptor on mouse lymphocytes and thymocytes in hypertonic medium, *Proc. Natl. Acad. Sci. U.S.A.,* 74, 4511, 1977.
40. **Yahara, I. and Kakimoto-Sameshima, F.,** Analysis of ligand-independent cap formation induced in hypertonic medium, *Exp. Cell Res.,* 119, 237, 1979.
41. **Schreiner, G. F., Braun, J., and Unanue, E. R.,** Spontaneous redistribution of surface immunoglobulin in the motile B lymphocytes, *J. Exp. Med.,* 144, 1683, 1977.
42. **Bourguignon, L. Y. W., Nagpal, M. L., and Hsing, Y. C.,** Phosphorylation of myosin light chain during capping of mouse T-lymphoma cells, *J. Cell Biol.,* 91, 889, 1981.
43. **Schrader, J. W., Cunningham, B. A., and Edelman, G. M.,** Functional interaction of viral and histocompatibility antigens at tumor cell surfaces, *Proc. Natl. Acad. Sci. U.S.A.,* 72, 5066, 1975.
44. **Abbas, A. K. and Unanue, E. R.,** Interrelationships of surface immunoglobulin and Fc receptors on mouse B-lymphocytes, *J. Immunol.,* 115, 1665, 1975.
45. **Forni, L. and Pernis, B.,** in *Membrane Receptors of Lymphocytes,* Seligmann, M., Pred'homme, J. L., and Kourilsky, F. M., Eds., North-Holland, Amsterdam, 1975, 193.
46. **Abrahansohn, I., Nilsson, U. R., and Abdou, N. I.,** Relationship of immunoglobulin to complement receptors of human B-cells, *J. Immunol.,* 112, 1931, 1974.
47. **Nussenzweig, V.,** Receptors for immune complexes on lymphocytes, *Adv. Immunol.,* 19, 217, 1974.
48. **Preud'Homme, J. L., Neauport-Sautes, C., Piat, S., Silvestre, D., and Kourilsky, F. M.,** Independence of HL-A antigens and immunoglobulin determinants on the surface of human lymphoid cells, *Eur. J. Immunol.,* 2, 297, 1972.
49. **Unanue, E. R., Dorf, M. E., David, C. S., and Benacerraf, B.,** The presence of I-receptor-associated antigens on B-cells in molecules distinct from immunoglobulin and H 2K and H 2D, *Proc. Natl. Acad. Sci. U.S.A.,* 71, 5014, 1974.
50. **Bourguignon, L. Y. W.,** Biochemical analysis of ligand-induced surface receptor patching and capping using a novel immunolactoperoxidase iodination technique, *J. Cell Biol.,* 83, 649, 1979.
51. **Majercik, M. H. and Bourguignon, L. Y. W.,** Insulin receptor capping and its correlation with Ca^{2+} and myosin light chain kinase, *J. Cell Biol.,* 99, 208a, 1984.
52. **Raz, A. and Bucana, C.,** The redistribution of membrane surface immunoglobulin induces the rearrangement of some membrane integral proteins, *Biochim. Biophys. Acta,* 597, 615, 1980.
53. **Bourguignon, L. Y. W. and Hsing, Y. C.,** The participation of adenylate cyclase in lymphocyte capping, *Biochim. Biophys. Acta,* 728, 186, 1983.
54. **Lin, D. C., Tobin, K. D., Grumet, M., and Lin, S.,** Cytochalasins inhibit nuclei-induced actin polymerization by blocking filament elongation, *J. Cell Biol.,* 84, 455, 1980.
55. **Bourguignon, L. Y. W. and Singer, S. J.,** Transmembrane interaction and the mechanism of capping of surface receptors by their specific ligands, *Proc. Natl. Acad. Sci. U.S.A.,* 74, 5031, 1977.
56. **Bourguignon, L. Y. W., Tokuyasu, K., and Singer, S. J.,** The capping of lymphocytes and other cells, studied by an improved method for immunofluorescence staining of frozen sections, *J. Cell Physiol.,* 95, 239, 1978.
57. **Butman, B. T., Bourguignon, G. J., and Bourguignon, L. Y. W.,** Lymphocyte capping induced by polycationized ferritin, *J. Cell Physiol.,* 105, 7, 1980.
58. **Laub, F., Kaplan, M., and Gitler, C.,** Actin polymerization accompanied Thy-1 capping on mouse thymocytes, *FEBS Lett.,* 124, 35, 1981.

59. **Yahara, I. and Edelman, G. M.,** The effects of concanavalin A on the mobility of lymphocyte surface receptors, *Exp. Cell Res.,* 81, 143, 1973.

60. **Lazarides, E.,** Intermediate filaments: a chemically heterogeneous developmentally regulated class of proteins, *Annu. Rev. Biochem.,* 51, 219, 1982.

61. **Steinert, P., Zackroff, R., Ayrardi-whitman, M., and Goldman, R.,** Isolation and characterization of intermediate filaments, *Meth. Cell Biol.,* 24, 399, 1982.

62. **Lazarides, E.,** Intermediate filaments as mechanical integrators of cellular space, *Nature (London),* 283, 249, 1980.

63. **Wang, E. and Choppin, P. W.,** Effect of vanadate on intracellular-distribution and function of 10 nm filaments, *Proc. Natl. Acad. Sci. U.S.A.,* 78, 2363, 1981.

64. **Howard, J., Eckert, B. S., and Bourguignon, L. Y. W.,** Comparison of cytoskeleton organization in canine distemper virus-infected and uninfected cells, *J. Gen. Virol.,* 64, 2379, 1983.

65. **Knapp, L. W., O'Guin, W. M., and Saeyer, R. H.,** Drug-induced alterations of cytokeratin organization in cultured epithelial cells, *Science,* 219, 501, 1983.

66. **Geiger, B. and Singer, S. J.,** Association of microtubules and intermediate filaments in chicken gizzard cells as detected by double immunofluorescence, *Proc. Natl. Acad. Sci. U.S.A.,* 77, 4769, 1980.

67. **Weber, K. and Osborn, M.,** in *Cytoskeletal Elements and Plasma Membrane Organization,* Poste, G. and Nicolson, G. L., Eds., Elsevier, Amsterdam, 1981, 2.

68. **Zackroff, R. V., Steinert, P., Whitman, M. A., and Goldman, R. D.,** in *Cytoskeletal Filaments and Plasma Membrane Organization,* Poste, G. and Nicolson, G. L., Eds., Elsevier, Amsterdam, 1981, 56.

69. **Geuens, G., de Brabander, M., Nuydens, R., and De May, J.,** The interaction between microtubules and intermediate filaments in cultured cells treated with taxol and nocodazole, *Cell Biol. Int. Rep.,* 7, 35, 1983.

70. **Zucker-Franklin, D.,** Capping and freeze-fracture analysis of Sezany cells, *Blood,* 54, 271, 1979.

71. **Bourguignon, L. Y. W. and Bourguignon, G. J.,** Immunological localization of intermediate filament proteins during lymphocyte capping, *Cell Biol. Int. Rep.,* 5, 783, 1981.

72. **Dellagi, K. and Brouet, J. C.,** Redistribution of intermediate filaments during capping of lymphocyte surface molecules, *Nature (London),* 298, 284, 1982.

73. **Lehto, V. P.,** 140,000 Dalton surface glycoprotein: a plasma membrane component of the detergent-resistant cytoskeletal preparations of cultured human fibroblasts, *Exp. Cell Res.,* 143, 271, 1983.

74. **Shay, J. W., Porter, K. R., and Prescott, D. M.,** The surface morphology and fine structure of CHO cells following enucleation, *Proc. Natl. Acad. Sci. U.S.A.,* 71, 3059, 1974.

75. **Shay, J. W., Gershenbaum, M. R., and Porter, K. R.,** Enucleation of CHO cells by means of cytochalasin B and centrifugation: the topography of enucleation, *Exp. Cell Res.,* 94, 47, 1975.

76. **Small, J. V. and Celis, J. E.,** Direct visulization of the 10 nm (100-A)-filament network in whole and enucleated cultured cells, *J. Cell Sci.,* 31, 393, 1978.

77. **Lehto, V. P., Virtanen, I., and Kurki, P.,** Intermediate filaments anchor the nuclei in nuclear monolayers of cultured human fibroblasts, *Nature (London),* 272, 175, 1978.

78. **Bennett, V.,** Purification of an active proteolytic fragment of the membrane attachment site, *J. Biol. Chem.,* 253, 2292, 1978.

79. **Weaver, D. C., Pasternack, G. R., and Marchesi, V. T.,** The structural basis of ankyrin function. II. Identification of two functional domains, *J. Biol. Chem.,* 259, 6170, 1984.

80. **Anderson, R. A. and Lovrien, R. E.,** Glycoprotein is linked by band 4.1 protein to the human erythrocyte membrane skeleton, *Nature (London),* 307, 655, 1984.

81. **Glenney, J. R. and Glenney, P.,** The microvillus 110 K cytoskeletal protein is an integral membrane protein, *Cell,* 37, 743, 1984.

82. **Glenney, J. R., Glenney, P., and Weber, K.,** F-actin binding and cross-linking properties of porcine brain fodrin, a spectrin-related molecule, *J. Biol. Chem.,* 257, 9781, 1982.

83. **Bennett, V., Davis, J., and Fowler, W. E.,** Brain spectrin, a membrane-associated protein related in structure and function to erythrocyte spectrin, *Nature (London),* 299, 126, 1982.

84. **Glenney, J. R., Glenney, P., and Weber, K.,** Erythroid spectrin, brain fodrin and intestinal brush border protein (TW 260/240) are related molecules containing a common calmodulin-binding subunit bound to a varient cell type-specific subunit, *Proc. Natl. Acad. Sci. U.S.A.,* 79, 4002, 1982.

85. **Branton, D., Cohen, C. M., and Tyler, J.,** Interaction of cytoskeletal proteins on the human erythrocyte membrane, *Cell,* 24, 24, 1981.

86. **Ungewickell, E., Bennett, P. M., Calvert, R., Ohanian, V., and Gratzer, N. B.,** *In vitro* formation of a complex between cytoskeletal proteins of the human erythrocyte, *Nature (London),* 280, 811, 1979.

87. **Ohanian, V., Wolfe, L. C., John, K. M., Pinder, V. C., Lux, S. E., and Gratzer, W. B.,** Analysis of the ternary interaction of the red cell membrane skeletal proteins, spectrin, actin and 4.1, *Biochemistry,* 23, 4416, 1984.

88. **Cohen, C. M. and Langley, R. C.,** Functional characterization of human erythrocyte spectrin alpha and beta chains: association with actin and erythrocyte protein 4.1, *Biochemistry,* 23, 4488, 1984.

89. **Anderson, R. A. and Lovrien, R. E.,** Glycophorin is linked by band 4.1 protein to the human erythrocyte membrane skeleton, *Nature (London),* 307, 655, 1984.

90. **Anderson, R. A. and Marchesi, V. T.,** Regulation of the association of membrane skeletal protein 4.1 with glycophorin by a polyphosphoinositide, *Nature (London),* 318, 295, 1985.

91. **Pasternak, G., Anderson, R. A., Leto, T. L., and Marchesi, V. T.,** Interactions between protein 4.1 and band 3: an alternative binding site for an element of the membrane skeleton, *J. Biol. Chem.,* 260, 3676, 1985.

92. **Granger, B. L. and Lazarides, E.,** Membrane skeletal protein 4.1 of avian erythrocytes is composed of multiple variants that exhibit tissue-specific expression, *Cell,* 37, 595, 1984.

93. **Goodman, S. R., Casoria, L., Coleman, D., and Zagon, I. S.,** Identification and location of brain protein 4.1, *Science,* 224, 1433, 1984.

94. **Baines, A. J. and Bennett, V.,** Synapsin I is a spectrin-binding protein immunologically related to erythrocyte protein 4.1, *Nature (London),* 315, 410, 1985.

95. **Bennett, V. and Stenbuck, P. J.,** Identification and partial purification of ankyrin, the high affinity membrane attachment site for human erythrocyte spectrin, *J. Biol. Chem.,* 254, 2533, 1979.

96. **Davis, J. and Bennett, V.,** Brain ankyrin: a membrane-associated protein with binding sites for spectrin, tubulin and the cytoplasmic domain of the erythrocyte anion channel, *J. Biol. Chem.,* 259, 13550, 1984.

97. **Bennett, V.,** Purification of an active proteolytic fragment of the membrane attachment site, *J. Biol. Chem.,* 253, 2292, 1978.

98. **Weaver, D. C. and Marchesi, V. T.,** The structural basis of ankyrin function. I. Identification of two functional domains, *J. Biol. Chem.,* 259, 6170, 1984.

99. **Weaver, D. C., Padternack, G. R., and Marchesi, V. T.,** The structural basis of ankyrin function. II. Identification of two functional domains, *J. Biol. Chem.,* 259, 6170, 1984.

100. **Bourguignon, L. Y. W., Suchard, S. J., Nagpal, M. L., and Glenney, J. R.,** A T-lymphoma trans-membrane glycoprotein (gp 180) is linked to the cytoskeletal protein, fodrin, *J. Cell Biol.,* 101, 477, 1985.

101. **Suchard, S. J. and Bourguignon, L. Y. W.,** Identification of a fodrin-containing transmembrane complex from mouse T-lymphoma cells, *J. Cell Biol.,* 101, 286a, 1985.

102. **Bourguignon, L. Y. W., Walker, G., Suchard, S. J., and Balazovich, K.,** A lymphoma plasma membrane-associated protein with ankyrin-like properties, *J. Cell Biol.,* 102, 2115, 1986.

103. **Fox, J. E. B. and Boyles, J. K.,** Platelets contain a membrane skeleton linked to plasma membrane glycoproteins through actin-binding protein, *J. Cell Biol.,* 101, 410a, 1985.

104. **Daniel, J. L., Holmsen, H., and Adelstein, R. S.,** Thrombin-stimulated myosin phosphorylation in intact platelets and its possible involvement in secretion, *Thromb. Haemostatis,* 38, 984, 1977.

105. **Bourguignon, L. Y. W., Field, S., and Bourguignon, G. J.,** Phosphorylation of a tropomyosin-like (30 KD) protein during platelet activation, *J. Cell Biochem.,* 29, 19, 1985.

106. **Bourguignon, L. Y. W., Walker, G., and Bourguignon, G. J.,** Phorbol ester-induced phosphorylation of a transmembrane glycoprotein (gp 180) in human blood platelets, *J. Biol. Chem.,* 260, 11775, 1985.

107. **Kwong, C. H. and Mueller, G. C.,** Antagonism of concanavalin A capping in phorbol ester-activated lymphocytes by calmodulin inhibitors and certain amino acid esters, *Cancer Res.,* 42, 2115, 1982.

108. **Williams, D. A., Boxer, L. A., Oliver, J. M., and Baehner, R. L.,** Cytoskeletal regulation of concanavalin A capping in pulmonary alveolar macrophages, *Nature (London),* 267, 255, 1977.

109. **Bourguignon, L. Y. W. and Kerrick, W. G. L.,** Receptor capping in mouse T-lymphocyte cells; a Ca^{2+} and calmodulin-stimulated ATP-dependent process, *J. Membrane Biol.,* 75, 65, 1983.

110. **Huxley, H. E.,** Molecular basis of contraction in cross-striated muscle, in *The Structure and Function of Muscle,* Bourne, G. H., Ed., Academic Press, New York, 1972, 302.

111. **Duncan, C. J.,** Ed., *Calcium in Biological Systems,* Cambridge University Press, London, 1976.

112. **Berridge, M. J.,** Inositol triphosphate and diacylglycerol as second messengers, *Biochem. J.,* 220, 345, 1984.

113. **Braun, J., Shaaf, R. I., and Unanue, E. R.,** Cross-linking by ligands to surface immunoglobulin triggers mobilization of intracellular Ca^{2+} in B-lymphocytes, *J. Cell Biol.,* 82, 755, 1979.

114. **Tsien, R. Y., Pozzan, T., and Rink, T. J.,** Calcium hemostasis in intact lymphocytes: cytoplasmic free calcium monitored with a new intracellular trapped fluorescent indicator, *J. Cell Biol.,* 94, 325, 1982.

115. **Pozzan, T., Arslan, P., Tsien, R. Y., and Rink, T. J.,** Antiimmunoglobulin, cytoplasmic free calcium, and capping in B lymphocytes, *J. Cell Biol.,* 94, 335, 1982.

116. **Rink, T. J., Smith, S. W., and Tsien, R. Y.,** Cytoplasmic free Ca^{2+} in human platelets: Ca^{2+} threshold and Ca^{2+} independent activation from shape-change and secretion, *FEB Lett.,* 148, 21, 1982.

117. **Rink, T. J., Sanchez, A., and Hallam, T. J.,** Diacylglycerol and phorbol ester stimulate secretion without raising cytoplasmic free calcium in human platelets, *Nature (London),* 305, 317, 1983.

118. **Hallam, T. J., Sanchez, A., and Rink, T. J.,** Stimulus-response coupling in human platelets: changes evoked by platelet-activating factor in cytoplasmic free calcium monitored with the fluorescent calcium indicator Quin, *Biochem. J.,* 218, 819, 1984.

119. **Sklar, L. A., Hyslop, P. A., Oades, Z. G., Omann, G. M., Jesaitis, A. J., Painter, R. G., and Cochrane, C. G.**, Signal transductions and ligand-receptor dynamics in the human neutrophil, *J. Biol. Chem.*, 260, 11461, 1985.

120. **Sklar, L. and Oades, Z. G.**, Signal transduction and ligand-receptor dynamics in the neutrophil, *J. Biol. Chem.*, 260, 11468, 1985.

121. **Sawyer, D. W., Sullivan, J. A., and Mandell, G. L.**, Intracellular free calcium localization in neutrophils during phagocytosis, *Science*, 230, 663, 1985.

122. **Haslam, R. J. and Davidson, M. M. L.**, Potentiation by thrombin of the secretion of serotonin from permeabilized platelets equilibrated with Ca^{2+} buffer, *Biochem. J.*, 222, 351, 1984.

123. **Cheung, W. Y.**, Calmodulin plays a pivital role in cellular regulation, *Science*, 207, 19, 1980.

124. **Kerrick, W. G. L. and Bourguignon, L. Y. W.**, Capping of mouse T-lymphoma cells is regulated by a calcium-activated myosin light chain kinase, *Proc. Natl. Acad. Sci. U.S.A.*, 81, 165, 1984.

125. **Levin, R. W. and Weiss, B.**, Binding of trifluoperazine to the calcium dependent activator of cyclic nucleotide phosphodiesterase, *Mol. Pharmacol.*, 13, 690, 1977.

126. **Bourguignon, L. Y. W. and Balazovich, K.**, Effect of antidepressant drug-staelazine on lymphocyte capping, *Cell Biol. Int. Rep.*, 4, 947, 1980.

127. **Nelson, G. A., Andrew, M. L., and Karnovsky, M. J.**, Participation of calmodulin in immunoglobulin capping, *J. Cell Biol.*, 95, 771, 1982.

128. **Cassidy, P., Hoar, P. E., and Kerrick, W. G. L.**, Irreversible thiophosphorylation and activation of tension in functional skinned rabbit ileum strips by [r-^{35}S]ATP S, *J. Biol. Chem.*, 254, 11148, 1980.

129. **Adelstein, R. S. and Klee, C. B.**, Smooth muscle myosin light chain kinase, in *Calcium and Cell Function*, Vol. 1, Cheung, W. Y., Ed., Academic Press, New York, 1980, 167.

130. **Dedman, J. R., Brinkley, B. R., and Means, A. R.**, Regulation of microfilaments and microtubules by calcium and cyclic AMP, *Adv. Cyclic Nucleotide Res.*, 11, 131, 1979.

131. **Adelstein, R. S. and Conti, M. A.**, Phosphorylation of platelet myosin increased actin-activated myosin ATPase activity, *Nature (London)*, 256, 597, 1975.

132. **Scordilis, S. P. and Adelstein, R. S.**, Myoblast myosin phosphorylation is prerequisite for actin-activation, *Nature (London)*, 268, 558, 1977.

133. **Bourguignon, L. Y. W., Nagpal, M. L., Balazovich, K., Guerriero, V., and Means, A. R.**, Association of myosin light chain kinase with lymphocyte membrane-cytoskeleton complex, *J. Cell Biol.*, 95, 793, 1982.

134. **Fay, F. S., Shlevin, H. H., Granger, W. C., and Taylor, S. R.**, Aequorin luminescence during activation of single isolated smooth muscle cells, *Nature (London)*, 280, 506, 1979.

Chapter 7

THE ROLE OF MEMBRANE POTENTIALS AND OF INTRACELLULAR pH IN SECRETORY CELL STIMULATION

Elizabeth R. Simons

TABLE OF CONTENTS

I. INTRODUCTION

All living cells examined to date maintain ionic gradients across their plasma membranes, in general including an inward Na^+ gradient (i.e., $[Na^+_{out}] \gg [Na^+_{in}]$), and an outward K^+ gradient ($[K^+_{in}] \gg [K^+_{out}]$), which are maintained by energy requiring pump mechanisms such as Na^+/K^+ ATPases, and others. These cells therefore maintain a resting membrane potential, usually negative inside with respect to the external medium, as well as an interior which is more acid than the extracellular environment. For excitable cells, and for those large enough to be impalable on microelectrodes, techniques such as patch clamping have permitted correlation between cellular responses and electrical signals. For smaller cells, as well as for those generally existing in suspension (rather than being attached to, or within, a cell layer), indirect measurements of electrical signals (e.g., changes in membrane potential) are now possible. A number of investigators have shown in the past few years that changes in the membrane potential, and in intracellular monovalent cation concentrations, accompany activation of such cells by many of the known stimuli for each respective cell type. The importance of these parameters has been extensively reviewed, in some cases with respect to a particular ion (e.g., the role of intracellular pH), in others with respect to specific cells.[1-14] Since the volume of literature is large and increasing, this chapter will concentrate on some of the current views on cation involvement in the stimulus-response of secretory cells such as human platelets[4,9,14-22] and neutrophils;[23-52] it should be noted that the final conclusions are not yet at hand and that disagreement in the published literature is still rampant.

It has been amply demonstrated[1-59] that exposure of platelets to specific agonists (such as thrombin) of neutrophils to soluble or particulate stimuli, as well as of lymphocytes or monocytes or fibroblasts to a variety of their specific stimuli, leads to a rapid, stimulus dose-dependent change in membrane potential as well as in cytoplasmic concentrations of Ca^{2+} and of H^+. The relative temporal relationships between membrane potential changes, Na^+ influxes, and alterations in intracellular Ca^{2+} and H^+ concentrations, i.e., their possible interdependence, have been the subject of intense investigation but have not yet been definitively determined. Similarly, it is just becoming clear that not only is none of the above changes sufficient by itself to initiate secretory cell activation[7,14,17] but, as will be discussed below, none appears to be completely necessary.[50] Nevertheless, for a given cell/stimulus pair for which a dose-dependent change in potential or in transmembrane cation gradients can be demonstrated, changes in these parameters can and are being used as indicators of cell stimulateability.[17,19,28,41,49,60,61] The cation gradient dependence of degranulation in secretory cells, as well as that of an oxidative burst in neutrophils, has also been examined.[9,17,25,29,31,41,48,52] Additionally, it has been shown that an important characteristic of the functional maturation of phagocytes (e.g., that of human bone marrow precursor cells to granulocytes,[62,63] of human myelocytic leukemia-derived HL-60 cells to granulocytes,[58,63,64] and of freshly isolated monocytes to macrophage-like cells[65]) is the development of a stimulus recognition capability and of an enhanced response to stimulation.

The sequence of events involved in phagocyte stimulation can be summarized as: (1) binding of a stimulus to its receptor/functional binding site, usually on or within the membrane; (2) alteration of the membrane permeability to specific cations and/or of binding of specific cations to membrane-associated proteins; (3) change in cytoplasmic concentration of cations; (4) activation of mediating entities (second messengers, phospholipases, proteases, enzymes of the oxidative burst, etc.); (5) for certain stimuli, motion up-gradient towards the source of chemotactic agonist; (6) for certain stimuli, formation and closure of phagovacuoles containing the stimulus-receptor complex; and (7) execution of the cell's function — degranulation and secretion of granule contents and oxidative products.[3,4,7,11] The exact sequence of these events remains unknown, but it seems likely that several may occur simultaneously.

Specifically, it has been shown that, for neutrophil stimulation, the initiating step, binding of certain stimuli to their specific receptors on the cell surface, induces a variety of responses. These include a rapid change in the cell's membrane potential, an equally, or even more, rapid change in cytoplasmic Ca^{2+} and H^+, the generation of a nonmitochondrial respiratory burst, degranulation, and changes in cell membrane and cytoskeletal characteristics that result in adherence, locomotion, phagocytosis, and/or aggregation.[7,11,65] The mechanism of information transduction by which the interaction of membrane receptors of these cells with their specific stimuli is transmitted to the reactive components of these cells (e.g., membrane-bound enzymes, ion channels, cytoplasmic entities, organelles) is thought to involve changes in cation gradients across the plasma membrane and in certain cation concentrations within the cytoplasm.[23,24,26] Similarly, it has been shown that stimulation of human platelets by specific agonists leads to a rapid membrane depolarization.[16,19,22] Examination in greater detail for one of the physiologically significant stimuli, thrombin, has demonstrated that thrombin-induced depolarization and degranulation of human platelets depend, respectively, on the Na^+ and K^+ gradients, and that the former is accompanied by an equally rapid and Na^+ influx-dependent cytoplasmic alkalinization.[16,17]

The coupling between stimulus-induced Na^+ and H^+ fluxes has been noted in many types of cells; in some the coupling is electrically neutral, i.e., one H^+ countertransported for one Na^+, denoting an antiport, while in others there is interdependence without the tight kinetic and quantitative coupling, a symport. The difference between these two types of Na^+/H^+ coupling is difficult to determine, as are the complex interactions when compartmentalization and the presence of organelles must be considered. The measurement of cation gradients, (e.g., of transmembrane potentials and of intracellular H^+ and Ca^{2+} concentrations), of their kinetics of change, and of their interdependence, is hence of great importance. This chapter will attempt to summarize the sometimes conflicting data available to date, utilizing as examples neutrophils and platelets, for which extensive literature exists.

II. MEASUREMENT OF THE TRANSMEMBRANE POTENTIAL OF CELLS IN SUSPENSION

The membrane potential can frequently be approximated by the constant field equation which, if Cl^- ions are distributed passively across the membrane, can be represented as a function of the membrane's permeabilities to Na^+ and K^+ and the relative concentrations of these cations inside and outside the cell by the Goldman-Hodgkin-Katz equation:[67,68]

$$\psi = RT \ln \frac{P(K^+)[K^+]_{out} + P(Na^+)[Na^+]_{out} + P(Cl^-)[Cl^-]_{in}}{P(K^+)[K^+]_{in} + P(Na^+)[Na^+]_{in} + P(Cl^-)[Cl^-]_{out}}$$

where P is the permeability for each cation through that particular cell membrane. Verification that the equation is obeyed is essential; it is usually accomplished by incubating the cells in question in isotonic media of varying K^+ concentrations and measuring changes in the potential when $P(K^+)$ is made much larger than those of other cations, i.e., when valinomycin is added to the system. Figure 1 shows typical curves for three types of circulating blood cells.

For relatively large and immobile cells, this potential can be measured directly by inserting a microelectrode into the cytoplasm through the plasma membrane, but for smaller cells in suspension indirect techniques are necessary. Most of these involve the membrane potential-dependent distribution of a lipophilic cation across the plasma membrane between the medium and the cell's cytoplasm, as well as across all the internal organellar membranes; one therefore measures an average potential. It is essential that the distribution be controlled by the

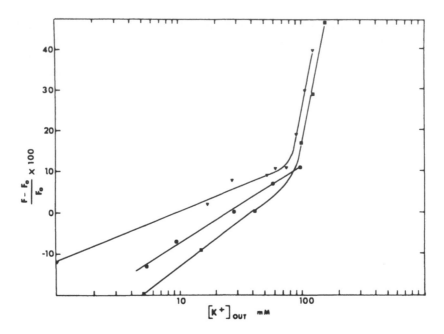

FIGURE 1.   Effect of external potassium concentration on the relative change of diSC$_3$(5) fluorescence upon addition of valinomycin. Valinomycin (2 × 10^6 M final concentration) was added to the following human blood cells equilibrated with 2 × 10^{-6} M diSC$_3$(5): (■——■) erythrocytes; (●——●) platelets; (▼——▼) granulocytes. (From Whitin, J. C. et al., *J. Biol. Chem.*, 255, 1874, 1980. With permission.)

parameter to be measured, and not by the probe concentration; the indicator distribution should be linearly proportional to the parameter being evaluated. The cation distribution is then detected by liquid scintillation counting of separated cells and their supernatant, if the probe is isotopically labeled, or by its absorbance or fluorescence, if these properties are detectable and proportional to the probe concentration.[67-77]

One technique is to use an isotopically labeled lipophilic cation such as the ^{3}H-triphenylmethylphosphonium or ^{3}H-tetraphenylphosphonium ion as an indicator of the resting membrane potential.[23,74,75] Slow changes in the potential can also be evaluated. However, since equilibration of these probes across the plasma membrane of mammalian cells tends to be very slow (10 to 120 min depending on the cell), their redistribution can also be slow and quantitative measurement of rapidly changing potentials in stimulus-responding secretory cells becomes difficult; in contrast, these probes equilibrate very rapidly across bacterial vesicles and have found great applicability in those studies.[74,75] This technique requires the rapid and effective separation of the cells from their external medium. One must also measure and correct for trapped external probe (using a nonpermeant isotopically labeled indicator such as labeled inulin or polyethyleneglycol) in order to evaluate the ratio of the extra- to intra-cellular probe concentrations which appears in the applicable version of the equation:

$$\psi = 59 \log[\text{TPP}^+]_{\text{out}}/[\text{TPP}^+]_{\text{in}}$$

Alternatively, one may use a fluorescent lipophilic cation as a membrane potential probe if its fluorescence can be shown to be a measure of its membrane potential-dependent distribution. Most of the probes currently in use are cyanine derivatives,[70,71] some of which have a great tendency to self-associate (with concomitant fluorescence quenching) while others do not. While the self-associating ones are the easiest to use for evaluations of relative membrane potentials or of membrane potential changes in suspensions, they cannot be used to investigate changes in single cells. Additionally, one cannot use their distribution to calculate the absolute membrane potential by means of the equation above since the total concentration of dye in the cells is generally unknown even though the external fluorescence is a linear function of the membrane potential. The great advantage of the optical probes is the short equilibration time required for their full distribution across the entire cell (including its organelles and granules) and their continuous monitorability.[35,67,69,71,73] While there are some disagreements on the ease with which absolute membrane potentials can be evaluated with these cyanine dyes it has been proven that, with adequate controls, they can be reliable indicators of relative cellular membrane potentials and of their changes.[16,24,26] Many (though not all) of the reported problems are attributable either to failure to verify one or more of the criteria described here or to excessively long preincubation of the cells with probe.[35,68-71,73] Since the probes are positively charged, excessive exposure may lead to slow discharge of the cells' resting potential and, in some cases, to poisoning of the mitochondria. Too low a concentration of probe can lead to artifacts, as the observed distribution ratio is then still a function of the limiting probe concentration rather than of the membrane potential being measured. Penetration into organelles (especially mitochondria) is unavoidable; luckily the initial stimulus response (<1 min) generally does not include release of this organellar content and can be shown (e.g. in platelets[15]) to involve the cytoplasm alone. Each laboratory must, however, verify the specific conditions it uses, the linearity of the probe's fluorescence response as a function of log $[K^+]_{out}$, and the cytoplasmic origin of the released probe, over the region of concentrations employed.

The available probes have been loosely separated into "fast" and "slow" responders,[71] the former exhibiting their signals in the micro- to millisecond range and therefore applicable to excitable cells, while the latter are slightly slower (response times in the millisecond to second range) but require less sophisticated instrumentation, have large signals, and are applicable to most nonexcitable cell stimulus-response coupling studies. The specific cyanine probe used is a matter of individual choice; of the many developed by Waggoner and his group, two have been the most popular and yield entirely comparable results: 3,3'-dipropylthiodicarbocyanine iodide (diSC$_3$(5)) and 3,3'-dipentyloxadicarbocyanine iodide (diOC$_5$(3)).[35,71] They are, respectively, of the self-associating and therefore intracellularly self-quenching, and of the nonassociating and nonself-quenching type. When data on mammalian cell membrane potentials obtained with these probes have been compared with those gathered with isotopically labeled probes such as triphenylmethylphosphonium (TPMP$^+$) or tetraphenylphosphonium (TPP$^+$)[74,75] ions or with microelectrodes, the results have been identical.[23-26,35,67,73] This is also true whether one utilizes diSC$_3$(5) or diOC$_5$(3), the choice depending on the particular type of study envisaged.[24,26] For cells in suspension, a self-associating and quenching probe like diSC$_3$(5) has great advantages (large signal, little light scattering interference since excitation and emission wavelengths are well separated, and, for diS cyanines, dual duty as a hydrogen peroxide indicator[28,61] as shown in Figure 2). In contrast, for cells which will be monitored individually (e.g., those being studied with a fluorescence-activated cell sorter), self-quenching would be a great detriment and probes of the oxacyanine type such as diOC$_5$(3) (Figure 3) must be used.[49,76,77] With either probe an identical rapid stimulus dose-dependent change in membrane potential (Figure 4) is observed.[24,26,35,63,73] An advantage inherent in using the fluorescent probes is that measurement of the stimulus-induced depolarization can be made rapidly and continuously, a capability which is required when one wishes to observe transient changes or rapidly changing systems since the requisite and time-consuming separation of cells from supernatant, or the long equilibration time necessary for the isotopically labeled TPMP$^+$ or TPP$^+$, is avoided.

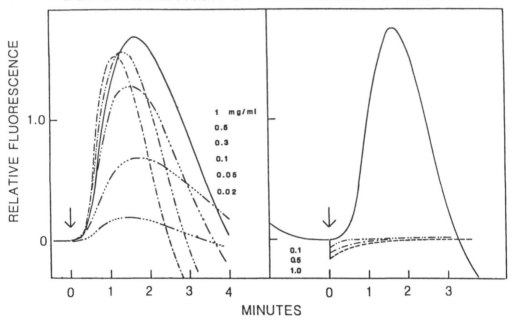

DEPOLARIZATION BY OPSONIZED ZYMOSAN

FIGURE 2. (Left) Effect of opsonized "clean" zymosan on fluorescence of $diSC_3(5)$-equilibrated neutrophils. (Right) Effect of unopsonized "clean" zymosan on fluorescence of $diSC_3(5)$-equilibrated neutrophils.

III. ROLE OF Na$^+$

We as well as others have used amiloride, or rapidly acting amiloride analogs, which block passive Na$^+$ channels and also affect passive H$^+$ channels (see, for example References 15, 17, 33, 36, 38, 51, 55, 57, 59, and 78). These agents, if left to incubate with the cells for several minutes, seriously perturb the cells' metabolic functions and ability to maintain normal gradients of cations other than Na$^+$. Care must therefore be taken to verify that other parameters of the resting cell, such as pH_i, and Ca^{2+}_i, are not affected by amiloride incubation for the desired time period. There are now, in addition to the original amiloride, a number of derivatives with somewhat altered properties;[17,51] the effect of the latter on resting cell potentials and cytoplasmic pH must hence be verified.

Since it is impossible to block all of the plasma membrane Na$^+$ entry mechanisms (ouabain, for example, inhibits only the Na$^+$/K$^+$ ATPase but none of the other pumps which might exist, such as Na$^+$/Ca^{2+}, and amiloride affects only passive Na$^+$ channels), one may also prevent Na$^+$ entry by substituting an isotonic choline buffer which maintains the normal K$^+$ gradient, but provides a noncell permeant and nonpumpable Na$^+$ analog without altering the resting potential of the cell, the latter being true at least for platelets[17] and for neutrophils.[25,31,41,45] In spite of this, there is some evidence that a choline buffer may alter pH_i after several minutes of incubation. One should therefore expose cells to this buffer very briefly before addition of the desired agonist.

By applying specific channel blockers (usually amilorides) and/or Na$^+$ depleted or replaced buffer systems, it has been shown that the major determinant for the resting membrane potential of platelets[17] and of neutrophils[25,29,41] is the external K$^+$ concentration. In contrast to platelets, whose initial response to stimulation (e.g., by thrombin) appears to be totally Na$^+$ controlled (blockable by amiloride, absent in choline/K$^+$ buffers), the stimulus response of neutrophils is only 50% Na$^+$-dependent.[17,25,41] It is, however, dependent upon the mode

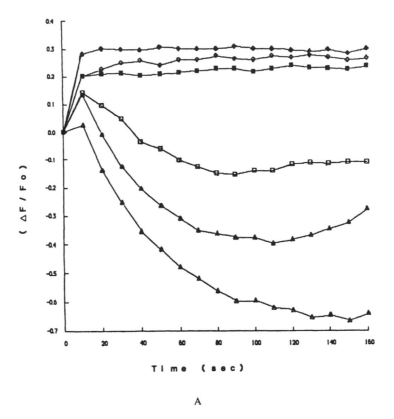

A

FIGURE 3.   (A) Changes in membrane potential in single cells stimulated with increasing doses of fMLP: $1 \times 10^{-11} M$ ($\blacklozenge$); $1 \times 10^{-10} M$ ($\lozenge$), $1 \times 10^{-9} M$ ($\blacksquare$); $1 \times 10^{-8} M$ ($\square$); $1 \times 10^{-7} M$ ($\blacktriangle$); and $1 \times 10^{-6} M$ ($\triangle$). These measurements were done simultaneously with indo-1 fluorescence measurements of changes in cytoplasmic Ca^{2+} concentration (see B) on the FACS. (B) Changes in cytoplasmic Ca^{2+} concentration measured with indo-1 in single cells stimulated with increasing doses of fMLP: same symbols as (A). These measurements were done simultaneously with $diOC_5(3)$ fluorescence measurements of changes in membrane potential (see A) on the FACS dual laser system. (From Lazarri, I. et al., *J. Biol. Chem.*, 261, 9710, 1986. With permission.)

of preparation of the neutrophils. Pfefferkorn[72] discovered that lymphocytes, prepared by the usual density gradient techniques, behaved differently when contaminating red cells were removed by brief hypotonic lysis (in distilled H_2O)[91] than by brief hypertonic lysis (in NH_4Cl).[31] Both are accepted and in many laboratories interchangeable according to personal preference techniques for ridding leukocytes, monocytes, or lymphocytes of contaminating erythrocytes. Yet the NH_4Cl-prepared cells exhibit a stimulus response requirement for extracellular Na^+ which is absent in H_2O-treated cells as well as a possibly perturbed intracellular pH.[72] Thus, the role of Na^+ in cellular activation processes is dependent on the cell, on its preparative history, as well as on the stimulus, on its dose, and on the conditions under which the activation occurs.

IV. ROLE OF K⁺

The importance of extracellular K^+ in stimulus response cannot be evaluated independently since no innocuous analog (one that does not also block other channels) for K^+ exists. While valinomycin is totally specific for K^+,[79] this ionophore also transports the ion across intra-

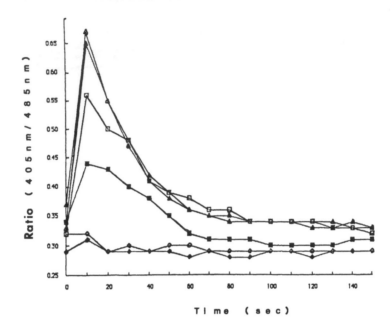

FIGURE 3B.

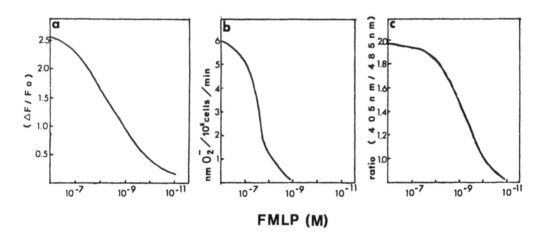

FMLP (M)

FIGURE 4.   Effect of fMLP on neutrophil membrane potential change, as indicated by the ratio $(F_{max} - F_0)/F_0$ of diSC$_3$(5) fluorescence (a); rate of superoxide formation, as measured by the rate of cytochrome c reduction (b); and changes in intracellular cytoplasmic [Ca^{2+}], as measured by the ratio of indo-1 fluorescence (excitation 355 nm) at 405/485 nm in cell suspensions of 10^6 cells/mℓ on a spectrofluorimeter (c). (From Lazarri, I. et al., *J. Biol. Chem.*, 261, 9710, 1986. With permission.)

cellular organellar membranes (see below), making interpretations difficult. The problem has therefore been approached in platelets,[17] and more recently in neutrophils,[25,41,52] by replacing extracellular Na$^+$ with K$^+$, an action which simultaneously reduces both the Na$^+$ and the K$^+$ gradient driving forces, and leads to depolarization of the cells. In this context, the actual contact time with these cell-perturbing buffers is important since all mammalian cells have strong homeostatic systems which lead them to attempt correction of this new state of affairs. Thus, the reason for conflicting reports on the importance of extracellular Na$^+$ and K$^+$ in eliciting maximal O$_2^-$ production upon activation of neutrophils with concanavalin A[25] or immune complexes[41] appears to be the length of incubation in high K$^+$/

low Na$^+$ or choline/K$^+$ buffers: short incubations elicit enhanced superoxide production,[41] while longer ones lead to inhibition.[25]

The K$^+$ gradient may also play an important role in the "end" effect of secretory cell function, granule content secretion. This is, for example, the controlling factor in platelet secondary granule secretion, a function which is totally independent of the Na$^+$ gradient.[17]

V. CYTOPLASMIC pH

It has been shown for many cells that stimulation is accompanied by changes in the cytoplasmic concentration of H$^+$ as well as that of Na$^+$, K$^+$, and Ca^{2+} ions;[2,6,7,15,41,45,46,81] both Na$^+$ and H$^+$ changes are frequently blocked by amiloride.[15,17,45,46,55] Although some distributive probes such as 9-amino acridine are applicable and yield the same results as those localized within the cell,[15] most of the current investigations utilize *in situ* probes, pH-sensitive fluorescein derivatives which are incubated with the cells in their nonfluorescent ester form to which the plasma membrane is freely permeable. Nonspecific esterases in the cellular cytoplasm then hydrolyze off the ester group, thus forming an *in situ* indicator which is pH sensitive and relatively membrane impermeable.

The original studies were performed with 5(6)-carboxyfluorescein, introduced into ascites cells[80] or platelets[15] as the diacetate ester. 5(6)-Carboxyfluorescein has an isosbestic point, a pH-independent absorbance at 464 nm, and a highly pH-dependent absorbance maximum at 492 nm, both giving rise to a fluorescence peak at 518 nm. The ratio mode is readily applicable to this system and has now been used for cell suspensions[15,80] as well as for fluorescence-activated cell sorter (FACS) studies.[82] The probe's main disadvantages are a relatively low pK, leading to a loss of pH sensitivity above pH 7.4 and some leakage in spite of its charge. A different fluorescein derivative, 2',7'-*bis*(carboxyethyl)-5(6)carboxyfluorescein (BCECF), exhibiting a higher pK (6.9) and pH sensitivity to 7.8 or 7.9 has recently become available.[83] It can be diffused into cells as the nonfluorescent acetoxymethyl ester (also subject to cytoplasmic esterase lysis); the de-esterified probe has a pH-sensitive fluorescence, and leaks less readily from cells, but has no isosbestic point.[33,36-40,46,81,84]

Calibration curves must be prepared for each probe/cell pair since the compartmental distribution, and therefore the indicated pH average for the cell, depends upon the cells themselves, the nature and the concentration of the probe, the cell purification process (see above), and the external buffer conditions. The preparation of these calibration curves was first described by Thomas et al. for ascites cells[80] and has been applied to other cells as well.[15]

Stimulus-induced changes in intracellular pH can be followed by using these fluorescein derivatives; they are rapid and dose dependent, as shown in Figure 5. Similar studies have been performed on platelets,[15,18] neutrophils,[37,39,40,44,46,47] lymphocytes,[38,42] fibroblasts,[57] and gastric glands[59] among others.

VI. IONOPHORES

It should be noted that such ionophores as valinomycin, nigericin, monensin, ionomycin, or A23187 are of limited utility in investigating the role of specific cations in cellular stimulus response. (1) All penetrate not only the plasma, but also internal organellar membranes, collapsing the gradients across these barriers, and causing the release of granular cations into the cytoplasm. These cations are not normally released into that space since degranulation is thought to occur at the outside of the plasma membrane surface into the extracellular milieu; thus, granules constitute a cation pool with which the cell itself does not usually deal. For example, the dense granules of platelets contain approximately 0.5 M Ca^{2+} while

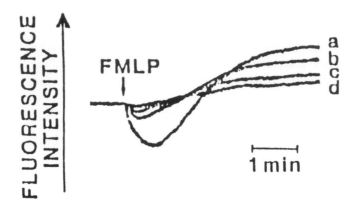

FIGURE 5. Effects of pertussis toxin on pHi changes in neutrophils induced by fMLP or A-23187. 100 n*M* fMLP were added to the cells in NaCl-medium which were pretreated with pertussis toxin (0.1 to 2.0 μg/ mℓ) at 37°C for 2 hours. (a) Control cells, (b)—(d) pertussis toxin-treated cells. The concentrations of pertussis toxin were as follows: (b) 0.2 μg/ mℓ, (c) 0.5 μg/mℓ, (d) 1.0 μg/mℓ. (From Satoh, M. et al., *Biochem. Biophys. Res. Commun.*, 131, 64, 1985. With permission.)

the cytoplasmic free Ca^{2+} is maintained at 100 n*M*, but none of the granule pool is involved in the transient activation-induced cytoplasmic Ca^{2+} increase.[4,85,86] (2) Except for valino-mycin, which solely transports K^+ down every existent transmembrane gradient, most ionophores are relatively unspecific.[79] Thus, nigericin has roughly equal affinities for K^+ and H^+, and 1/100th that affinity for Na^+; in view of the high concentrations of K^+ inside and Na^+ outside most blood cells, nigericin therefore effectively collapses K^+, H^+, and Na^+ gradients and little can be deduced about the involvement of any single one of these three ions from the use of nigericin alone. Similarly, monensin has equal affinities for Na^+ and H^+, and 1/10th that affinity for K^+. A23187, a classic "Ca^{2+}" ionophore, transports two H^+ for every Ca^{2+} and therefore affects intracellular and intraorganellar pH as well as Ca^{2+} concentrations.[79] One must therefore use combinations of blocking agents and/or altered extracellular cation concentrations, in preference to ionophores, in order to decipher the dependence of cellular response on each of these cations. The ionophores are, however, useful as positive controls to assure that the resting cells do indeed maintain the appropriate cation gradients.

VII. NEUTROPHILS

Neutrophils can be activated via different mechanisms which all lead to phagocytosis, degranulation, and the formation of toxic oxidative entities such as superoxide, peroxide, oxygen and hydroxyl free radicals, and, in the presence of halide, hypochlorite anions. Certain stimuli act through specific neutrophil membrane receptors, for example the che-moattractant peptides (such as C3a, C5a, certain products of stimulated leukocytes, and some synthetic peptides of the formyl-methionyl-leucyl-phenylalanine type[85,86]), the Fc por-tion of immunoglobulin (whether the antibody be deposited on immune complexes, deposited on a foreign surface, or as an immunoglobulin aggregate[87,88]), and the C3b complement fragment covalently attached to serum opsonized particles.[89,90]

Binding of the stimulus generally induces a rapid change in the membrane potential of all phagocytes, a change which has been extensively documented in a number of laboratories. It has now been reported that neutrophils undergo a dose-dependent depolarization with phorbol esters,[23,24] opsonized particles,[32] chemotactic factors,[85,86] and immune complexes.[23]

These phenomena appear to be independent of species since they have been reported for human, rabbit, guinea pig, and rat phagocytes. A biphasic response to formyl-methionyl-leucyl-phenylalanine (fMLP) by neutrophils[49] and by myeloid precursor cells[62] has been noted and will be discussed below.

The cation gradient changes associated with neutrophil activation have been extensively reviewed.[6,7,10,11,25] The importance of transmembrane Na^+ and K^+ gradients, as well as of the presence of these ions, per se, in the extracellular buffer, has not been fully delineated as conflicting conclusions still exist.[23,25,41] These are made even more complex by the previously mentioned discovery by Pfefferkorn[72] that a single minor step in neutrophil preparation, the lysis of residual erythrocytes in the Ficoll-Hypaque isolate, can alter Na^+ requirements: when the lysis is accomplished with ice cold H_2O, extracellular Na^+ is not necessary for oxidative burst expression, but when lysis is accomplished by hyper- or hypotonic lysis with NH_4Cl, external Na^+ becomes an absolute requirement, an observation which we have recently confirmed.[41] The difference appears to lie in the NH_4Cl attributable irreversible change in the internal pH.[72] Unfortunately, both procedures are accepted and have a long history of use in neutrophil preparation — indeed are used interchangeably in some laboratories.

The monovalent cation gradient involvement is also complicated by the highly active pump system in neutrophils which causes them to respond to, and attempt to correct for, changes in their extracellular milieu. The investigations are usually performed by utilizing buffers which alter the Na^+ and K^+ gradients and extracellular availability,[6,7,25,41] substituting K^+ for Na^+ salts when reduction both in gradients and in membrane potential is desired, and choline for Na^+ when the membrane potential and the transmembrane K^+ gradients are to be left intact. However the cells' homeostatic mechanism is rapid and the results obtained are, in part, dependent on the time of exposure to the unphysiological buffer systems. For example, we have found that Na^+ and K^+ ions are each responsible for approximately 50% of the immune complex-induced neutrophil depolarization when exposure to these buffers is short (2 min), and that superoxide generation is greatly enhanced when extracellular Na^+ is virtually absent,[41] while Korchak and Weissmann report it to be inhibited when cells were preincubated for 10 min, and found that extracellular Na^+ is required for depolarization.[25] It should be noted (although this chapter will not deal with Ca^{2+} involvement) that there is also general disagreement on stimulation requirements for extracellular Ca^{2+} which appears to be similarly attributable to the varying incubation times employed by different researchers. There are other reports of Na^+- or K^+-dependent neutrophil activation steps: an increase in fMLP affinity has been reported[31,48] when Na^+ is replaced by K^+, choline, or sucrose (i.e., whether or not the cells are depolarized), indicating an affinity enhanced by lowered $[Na^+]_{out}$ as has an enhanced lectin susceptibility when $[K^+]_{out}$ is increased, a K^+ specific augmented capping since substitution of choline for Na^+ produces no such enhancement.[41,52]

Thus, the net effect of neutrophil activation on its membrane potential, approximately -70 mV in resting cells, depends upon the nature and concentration of the stimulus and upon the ionic environment in which the stimulation occurs. Neutrophils suspended in physiological buffers (0.145 to $0.150 M Na^+$) respond rapidly, the time and extent of maximal change being dependent on the specific stimulus and on its dose.[24,26] Generally, a depolarization has been observed at saturating and even subsaturating concentrations of chemotactic peptides, phorbol esters, immune complexes, and opsonized particles, whether measured on suspensions of cells (e.g., in a fluorimeter)[23,24,26,27,32] or on individual ones (e.g., in a flow cytometer).[49,76,77] A hyperpolarization, first noted by Korchak and Weissmann,[23] has been reconfirmed.[49] By using flow cytometry, we have detected a very rapid hyperpolarization, maximal at 10 to 12 sec and constant thereafter, in response to very low ($10^{-11} M$, i.e. threshold), doses of fMLP (Figure 3);[49] as the dose is increased to 10^{-10} and 10^{-9}, the extent of hyperpolarization decreases and, by $10^{-8} M$ fMLP, a later depolarization ($t > 30$

sec) begins to appear. At saturating doses of fMLP (above 10^{-7}), the hyperpolarization is barely expressed and only the depolarization is detectable. These results, more readily obtained in a flow cytometer, where one observes the cells themselves, than in a fluorimeter, where one observes supernatant (a large volume) plus cells (a much smaller internal volume), imply a still more complicated activation mechanism than had been suspected. In immature cells, the hyperpolarization described for neutrophils exposed to marginally effective stimulation (10^{-11} M fMLP) is observed in response to 10^{-7} M fMLP, saturating for mature neutrophils, when the cells are undifferentiated; as they mature to granulocytes, after treatment with GMCSF (granulocyte-monocyte colony stimulating factor), there is a differentiation-stage dependent decrease of hyperpolarization and, eventually, a depolarization similar to that described above for granulocytes.[62] The reasons for this low dose-dependent (for neutrophils) and differentiation-stage dependent (for the precursor cells) hyperpolarization have not yet been delineated. A very rapid rise in cytosolic Ca^{2+} concentrations is also part of the phagocytic cell's response to stimulation,[56] but will be discussed elsewhere in this volume.

The other monovalent cation of importance in neutrophil responses is the proton, H^+. A rapid change in cytoplasmic pH accompanies stimulus-induced membrane potential changes in many cellular systems.[1,2,6,7] In contrast to platelets, whose cytoplasm becomes more alkaline upon stimulation,[15] neutrophils initially become more acid (although realkalinization begins after less than 30 sec) as first reported by Grinstein,[36] and confirmed by several other groups[33,47] investigating the responses after very short time intervals (seconds), but difficult to visualize when data are collected only after 1 min.[46,81] There is agreement that the alkalinization begins 30 to 60 sec after stimulation, and is inhibited by amiloride,[81] and that the preceding acidification is unaffected by this blocker of passive Na^+ channels. Therefore, the initial stimulus-induced neutrophil response does not appear to proceed via a Na^+/H^+ antiport although a later step in the activation process may involve such countertransport. Evidence for human[45,81] and for rabbit[29] neutrophils points to the presence of such an antiport, although disagreement exists on whether it takes part in the initial stage or in a secondary event in the neutrophil response. In this sense, as has also been true for platelets,[15] lymphocytes,[83] and fibroblasts,[55] the presence of an antiport has been inferred from the amiloride blockage of both Na^+ and H^+ concentration changes. Simchowitz[45,46,81] has calculated the rates of the amiloride-inhibitable proton efflux and Na^+ influx during the alkalinization of the cytoplasm as being 15 and 17 meq/mℓ/min, verifying the 1:1 stoichiometry, the presence of a true antiport. In none of the other cases, however, has the stoichiometry and the time course been reported; the distinction between a true antiport with 1:1 stoichiometry and a channel open to both ions has hence not yet been made for lymphocyte and fibroblast activation.

VIII. PLATELETS

The role of cation gradients in platelet activation has recently been reviewed.[14] Like other secretory cells, platelets exhibit a negative resting membrane potential (approximately -50 mV[15,21]). Their recognition of a specific stimulus like thrombin with a rapid dose-dependent depolarization[15,16] is accompanied by an equally rapid alkalinization of the cytoplasm and transient influx of Na^+, and is followed by a sequential and energy-dependent secretion of the contents of their dense, alpha, and secondary (lysosomal) granules, in that order.[4,92] The dependence upon individual monovalent cations differs, however, from that exhibited by neutrophils. For example, although the resting potential is unaltered when Na^+ is replaced by choline, implying that it is attributable largely to the K^+ gradient across the plasma membrane, collapse of that gradient with valinomycin has no effect whatsoever on thrombin-induced depolarization; such a procedure does, however, prevent degranulation (granule content secretion). In contrast, blockage of the previously mentioned stimulus-elicited Na^+ influx with amiloride totally abrogates the platelet depolarization and the accompanying cytoplasmic alkalinization, but the degranulation proceeds normally. Thus, unlike the neutrophil, in which stimulus responses appear to depend upon both Na^+ and K^+ gradients, the corresponding responses in platelets involve only one of these cations — and a different one for depolarization and for degranulation. There is a concomitant rapid rise in cytoplasmic Ca^{2+}, which, like the corresponding one for neutrophils, will be discussed elsewhere in these volumes.

In summary, the current level of understanding of stimulus-response coupling in secretory cells is that, depending on the specific cell, the existence of Na^+ and K^+ gradients is vital for a normal response. In most of these cells a Na^+/H^+ antiport or symport is involved, though not necessarily in the initial step.

ACKNOWLEDGMENTS

Experiments in this laboratory were made possibly in part by National Institutes of Health grants HL15335, HL07501, AM35106, and HL33565, whose support is gratefully acknowledged.

REFERENCES

1. **Busa, W. B.**, Metabolic regulation via intracellular pH, *Am. J. Physiol.*, 246, R409, 1984.
2. **Busa, W. B.**, Mechanism and consequences of pH-mediated cell regulation, *Annu. Rev. Physiol.*, 48, 389, 1986.
3. **Becker, E. L.**, Leukocyte stimulation: receptor, membrane and metabolic events, *Fed. Proc.*, 45, 2148, 1986.
4. **Holmsen, H.**, Platelet metabolism and activation, *Semin. Hematol.*, 22, 219, 1985.
5. **Roos, A. and Boron, W. F.**, Intracellular pH, *Physiol. Rev.*, 61, 296, 1981.
6. **Sha'afi, R. I. and Naccache, P. H.**, Ionic events in neutrophil chemotaxis and secretion, *Adv. Inflammation Res.*, 2, 115, 1981.
7. **Simchowitz, L.**, Current trends in leukocyte activation, in *The Year in Immunology, 1984—85*, Cruse, J. M. and Lewis, Jr. E. R., S. Karger, Basel, 1985, 122.
8. **Nuccitelli, R. and Deamer, D. W., Eds.**, *Intracellular pH: Its Measurement, Regulation, and Utilization in Cellular Functions*, Alan R. Liss, New York, 1982.
9. **Lages, B., Scrutton, M. C., and Holmsen, H.**, Secretion by gel-filtered human platelets: response of platelet Ca^{2+}, Mg^{2+}, and K^+ to secretory agents, *J. Lab. Clin. Med.*, 90, 873, 1977.
10. **Cochrane, C. G.**, Mechanisms coupling stimulation and function of leukocytes, *Fed. Proc.*, 43, 2729, 1984.
11. **Korchak, H. M., Vienne, K., Rutherford, L. E., and Weissman, G.**, Neutrophil stimulation: receptor, membrane, and metabolic events, *Fed. Proc.*, 45, 2749, 1986.
12. **Stossel, T., Weissman, G., and Cochrane, C.**, Mechanisms coupling stimulation and function in leukocytes, *Fed. Proc.*, 43, 2729, 1984.
13. **Simchowitz, L. and Roos, A.**, Regulation of intracellular pH in human neutrophils, *J. Gen. Physiol.*, 85, 442, 1985.
14 **Simons, E. R. and Greenberg-Sepersky, S. M.**, Transmembrane monovalent cation gradients, *Platelet Responses and Metabolism*, Vol. 3, Holmsen, H., Ed., CRC Press, Boca Raton, Fla., 1987, 31.
15. **Horne, W. C., Norman, N. E., Schwartz, D. B., and Simons, E. R.**, Changes in cytoplasmic pH and in membrane potential in thrombin-stimulated human platelets, *Eur. J. Biochem.*, 120, 295, 1981.
16. **Horne, W. C. and Simons, E. R.**, Probes of transmembrane potentials in platelets: changes in cyanine dye fluorescence in response to aggregation stimuli, *Blood*, 51, 741, 1978.
17. **Greenberg-Sepersky, S. M. and Simons, E. R.**, Cation gradient dependence of the steps in thrombin stimulation of human platelets, *J. Biol. Chem.*, 259, 1502, 1984.
18. **Simons, E. R., Schwartz, D. B., and Norman, N. E.**, Stimulus response coupling in human platelets: thrombin-induced changes in pH$_i$, in *Intracellular pH: Its Measurement, Regulation, and Utilization in Cellular Functions*, Nuccitelli, R. and Deamer, D., Eds., Alan R. Liss, New York, 1982, 463.

19. **Pipili, E.**, Platelet membrane potential: simultaneous measurement of diSC$_{3(5)}$ fluorescence and optical density, *Thrombosis Haemostasis*, 54, 645, 1985.

20. **Friedhoff, L. T., Kim, E., Priddle, M., and Sonenberg, M.**, The effect of altered transmembrane ion gradients on membrane potential and aggregation of human platelets in blood plasma, *Biochem. Biophys. Res. Commun.*, 102, 832, 1981.

21. **Friedhoff, L. T. and Sonenberg, M.**, The membrane potential of human platelets, *Blood*, 61, 180, 1983.

22. **MacIntyre, D. E. and Rink, T. J.**, The role of platelet membrane potential in the initiation of platelet aggregation, *Thrombosis Haemostasis*, 47, 22, 1982.

23. **Korchak, H. M. and Weissmann, G.**, Changes in membrane potential of human granulocytes antecede the metabolic responses to surface stimulation, *Proc. Natl. Acad. Sci. U.S.A.*, 75, 3818, 1978.

24. **Whitin, J. C., Chapman, C. E., Simons, E. R., Chovaniec, M. E., and Cohen, H. J.**, Correlation between membrane potential changes and superoxide production in human granulocytes stimulated by phorbol myristate acetate: evidence for defective activation in chronic granulomatous disease, *J. Biol. Chem.*, 255, 1874, 1980.

25. **Korchak, H. M. and Weissmann, G.**, Stimulus-response coupling in the human neutrophil transmembrane potential and the role of extracellular Na$^+$, *Biochim. Biophys. Acta*, 601, 180, 1980.

26. **Seligman, B. E. and Gallin, J. I.**, Use of lipophilic probes of membrane potential to assess human neutrophil activation, *J. Clin. Invest.*, 66, 493, 1980.

27. **Seligman, B. E., Gallin, E. K., Martin, D. L., Shain, W., and Gallin, J. I.**, Interaction of chemotactic factors with human polymorphonuclear leukocytes: studies using a membrane potential sensitive cyanine dye, *J. Membrane Biol.*, 52, 257, 1980.

28. **Whitin, J. C., Clark, R. A., Simons, E. R., and Cohen, H. J.**, Effects of the myeloperoxidase system on fluorescent probes of granulocyte membrane potential, *J. Biol. Chem.*, 256, 8904, 1981.

29. **Naccache, P. H., Showell, H. J., Becker, E. L., and Sha'afi, R. I.**, Transport of sodium, potassium, and calcium across rabbit polymorphonuclear leukocyte membranes, *J. Cell Biol.*, 73, 428, 1977.

30. **Jones, G. S., Van Dyke, K., and Castranova, V.**, Transmembrane potential changes associated with superoxide release from human granulocytes, *Biochim. Biophys. Acta*, 645, 49, 1981.

31. **Simchowitz, L., Spilberg, I., and De Weer, P.**, Sodium and potassium fluxes and membrane potential of human neutrophils, *J. Gen. Physiol.*, 79, 453, 1982.

32. **Cohen, H. J., Newburger, P. E., Chovaniec, M. E., Whitin, J. C., and Simons, E. R.**, Opsonized zymosan-stimulated granulocytes-activation and activity of the superoxide-generating system and membrane potential changes, *Blood*, 58, 975, 1981.

33. **Della Bianca, V., Bellavita, P., DeTogni, P., Fumarulo, R., and Rossi, F.**, Studies on stimulus-response coupling in human neutrophils. I. Role of monovalent cations in the respiratory and secretory response to N-formyl-methinyl-leucyl-phenylalanine, *Biochim. Biophys. Acta*, 755, 497, 1983.

34. **De Tongi, P., Della Bianca, V., Bellavite, P., Grzeckowiak, M., and Rossi, F.**, Studies on stimulus-response coupling in human neutrophils. II. Relationships between the effects of changes of external ionic composition on the properties of N-formylmethionylleucylphenylalanine receptors and on respiratory and secretory responses, *Biochim. Biophys. Acta*, 755, 506, 1983.

35. **Seligman, B. E. and Gallin, J. I.**, Comparison of indirect probes of membrane potential utilized in studies of human neutrophils, *J. Cell Physiol.*, 115, 105, 1983.

36. **Grinstein, S. and Furuya, W.**, Amiloride-sensitive Na$^+$/K$^+$ exchange in human neutrophils: mechanism of activation by chemotactic factors, *Biochem. Biophys. Res. Commun.*, 122, 755, 1984.

37. **Gabig, T. G., Lefker, B. A., Ossanna, P. J., and Weiss, S. J.**, Proton stoichiometrey associated with human neutrophil respiratory-burst reactions, *J. Biol. Chem.*, 259, 13166, 1984.

38. **Grinstein, S., Cohen, S., Goetz, J. D., and Rothstein, A.**, Osmotic and phorbol ester-induced activation of Na$^+$/H$^+$ exchange: possible role of protein phosphorylation in lymphocyte volume regulation, *J. Cell Biol.*, 101, 269, 1985.

39. **Grinstein, S., Elder, B., and Furuya, W.**, Phorbol ester-induced changes of cytoplasmic pH in neutrophils: role of exocytosis in Na$^+$-H$^+$ exchange, *Am. J. Physiol.*, 248, C379, 1985.

40. **Grinstein, S., Furuya, W., and Biggar, W. D.**, Cytoplasmic pH regulation in normal and abnormal neutrophils, *J. Biol. Chem.*, 261, 512, 1986.

41. **Luscinskas, F. W., Mark, D. E., Lionetti, F. J., Cragoe, E. J., and Simons, E. R.**, The role of transmembrane cationic gradients in immune complex stimulation of human neutrophils, *Biophys. J.*, 47(abstr.), 201A, 1985.

42. **Mills, G. B., Cragoe, E. J., Jr., Gelfand, E. W., and Grinstein, S.**, Interleukin 2 induces a rapid increase in intracellular pH through activation of a Na$^+$/H$^+$ antiport, *J. Biol. Chem.*, 260, 12500, 1985.

43. **Molski, T. F. P., Naccache, M., Volpi, L. M., Wolpert, L. M., and Sha'afi, R. I.**, Specific modulation of the intracellular pH of rabbit neutrophils by chemotactic factors, *Biochem. Biophys. Res. Commun.*, 94, 508, 1980.

44. **Mollinedo, F., Manara, F. S., and Schneider, L. L.**, Acidification activity of human neutrophils, *J. Biol. Chem.*, 261, 1077, 1986.

45. **Simchowitz, L.,** Chemotactic factor-induced activation of Na^+/H^+ exchange in human neutrophils, I. Sodium fluxes, *J. Biol. Chem.*, 260, 13237, 1985.

46. **Simchowitz, L.,** Intracellular pH modulates the generation of superoxide radicals by human neutrophils, *J. Clin. Invest.*, 76, 1079, 1985.

47. **Satoh, M., Nanri, H., Takeshige, K., and Minakami, S.,** Pertussis toxin inhibits intracellular pH changes in human neutrophils stimulated by N-formylmethionyl-leucyl-phenylalanine, *Biochem. Biophys. Res. Commun.*, 131, 64, 1985.

48. **Zigmond, S. H., Woodworth, A., and Daukas, G.,** Effects of sodium on chemotactic peptide binding to polymorphonuclear leukocytes, *J. Immunol.*, 135, 531, 1985.

49. **Lazzari, K. G., Proto, P. J., and Simons, E. R.,** Simultaneous measurement of stimulus-induced changes in cytoplasmic Ca^{2+} and in membrane potential of human neutrophils, *J. Biol. Chem.*, 261, 9710, 1986.

50. **Lyman, C., Simons, E. R., Melnick, D. A., and Diamond, R. D.,** Neutrophil activation by unopsonized *candida* hyphae, submitted.

51. **Simchowitz, L. and Cragoe, E. J., Jr.,** Regulation of human neutrophil chemotaxis by intracellular pH, *J. Biol. Chem.*, 261, 6492, 1986.

52. **Roberts, R. L., Mounessa, N. L., and Gallin, J. I.,** Increasing extracellular potassium causes calcium-dependent shape change and facilitates concanavalin A capping in human neutrophils, *J. Immunol.*, 132, 2000, 1984.

53. **Deutsch, C. J., Holian, A., Holian, S. K., Daniele, R. P., and Wilson, D. F.,** Transmembrane electrical and pH gradients across human erythrocytes and human peripheral lymphocytes, *J. Cell. Physiol.*, 99, 79, 1979.

54. **Larsen, N. E., Enelow, R. I., Simons, E. R., and Sullivan, R.,** Effects of bacterial endotoxin on the transmembrane electrical potential and plasma membrane fluidity of human monocytes, *Biochim. Biophys. Acta*, 815, 1, 1985.

55. **L'Allemain, G., Franchi, A., Cargoe, E., Jr., and Pouyssegur, J.,** Blockade of the Na^+/H^+ antiport abolishes growth factor-induced DNA synthesis in fibroblasts, *J. Biol. Chem.*, 259, 4313, 1984.

56. **Tsien, R. Y., Pozzan, T., and Rink, T. J.,** T-cell mitogens cause early changes in cytoplasmic free Ca^{2+} and membrane potential in lymphocytes, *Nature (London)*, 295, 68, 1982.

57. **Moolenaar, W. H., Tsien, R. Y., Van der Saag, P. T., and de Laat, S. W.,** Na^+/H^+ exchanges cytoplasmic pH in the action of growth factors in human fibroblasts, *Nature (London)*, 304, 1985.

58. **Naccache, P. H., Molski, T. F. P., Spinelli, B., Borgeat, P., and Abboud, C. N.,** Development of calcium and secretory responses in the human promyelocytic leukemia cell line HL-60, *J. Cell. Physiol.*, 119, 241, 1984.

59. **Paradiso, A. M., Tsien, R. Y., and Machen, T. E.,** Na^+-H^+ exchange in gastric glands as measured with a cytoplasmic-trapped, fluorescent pH indicator, *Proc. Natl. Acad. Sci. U.S.A.*, 81, 7436, 1984.

60. **Tauber, A. I., Borregaard, N., Simons, E. R., and Wright, J.,** Chronic granulomatous disease: a syndrome of phagocyte oxidase deficiencies, *Medicine (Baltimore)*, 62, 286, 1983.

61. **Pagonis, C., Tauber, A. I., Pavlotsky, N., and Simons, E. R.,** Flavonoid impairment of neutrophil response, *Biochem. Pharmacol.*, 35, 237, 1986.

62. **Sullivan, R., Melnick, D. A., Meshulam, T., Simons, E. R., Lazzari, K. G., Proto, P., and Griffin, J. D.,** The effects of phorbol myristate acetate and chemotactic peptide on transmembrane potentials and cytosolic free calcium in mature granulocytes are absent in normal granulocyte precursors and develop as the cells mature, submitted.

63. **Kitagawa, S., Ohta, M., Nojiri, H., Kakinuma, K., Saito, M., Takaku, F., and Miura, Y.,** Functional maturation of membrane potential changes and superoxide-producing capacity during differentiation of human granulocytes, *J. Clin. Invest.*, 73, 1062, 1984.

64. **Newburger, P. E., Speier, C., Borregaard, N., Walsh, C. E., Whitin, J. C., and Simons, E. R.,** Development of the superoxide-generating system during differentiation of the HL-60 human promyelocytic leukemia cell, *J. Biol. Chem.*, 259, 3771, 1984.

65. **Bernardo, J., Brink, H. F., and Simons, E. R.,** Time dependence of transmembrane potential changes and intracellular calcium flux in stimulated human monocytes, *Fed. Proc.*, 45, 852, 1986.

66. **Weissman, G., Smolen, J., Korchak, H., and Hoffstein, S.,** The secretory code of the neutrophil, in *Cellular Interactions*, Dingle, J. T. and Gordon, J. L., Eds., Elsevier, Amsterdam, 1981, 15.

67. **Hoffman, J. F. and Laris, P. C.,** Determinations of membrane potentials in human and amphiuma red blood cells by means of a fluorescent probe, *J. Physiol. (London)*, 239, 519, 1974.

68. **Freedman, J. C. and Hoffman, J. F.,** Ionic and osmotic equilibrium of human red blood cells treated with nystatin, *J. Gen. Physiol.*, 74, 187, 1979.

69. **Freedman, J. C. and Laris, P. C.,** Electrophysiology of cells and organelles: studies with optical potentiometric indicators, *Int. Rev. Cytol.*, 12, 177, 1981.

70. **Sims, P. J., Waggoner, A. S., Wang, C. H., and Hoffman, J. F.,** Studies of the mechanisms by which cyanine dyes measure membrane potential in red blood cells and phosphatidyl choline vesicles, *Biochemistry*, 13, 3315, 1974.

71. **Waggoner, A. S.,** Dye indicators of membrane potential, *Annu. Rev. Biophys. Bioeng.,* 8, 47, 1979.
72. **Pfefferkorn, L. C.,** Transmembrane signaling: an ion flux independent model for signal transduction by complexed Fc receptors, *J. Cell Biol.,* 99, 2231, 1984.
73. **Bashford, C. L., Alder, C. M., Gray, M. A., Micklem, K. J., Taylor, C. C., Turek, P. J., and Pasternak, C. A.,** Oxonol dyes as monitors of membrane potential: the effect of viruses and toxins on the plasma membrane potential of animal cells in monolayer culture and in suspension, *J. Cell. Physiol.,* 123, 326, 1985.
74. **Ramos, S. and Kaback, H. R.,** The electrochemical proton gradient in *Escherichia coli* membrane vesicles, *Biochemistry,* 16, 848, 1977.
75. **Lichtstein, D., Kaback, H. R., and Blume, A. J.,** Use of a lipophilic cation for determination of membrane potential in neuroblastoma-glioma thybid cell suspensions, *Proc. Natl. Acad. Sci. U.S.A.,* 76, 650, 1979.
76. **Seligmann, B., Chused, T. M., and Gallin, J. I.,** Human neutrophil heterogeneity identified using flow microfluorometry to monitor membrane potential, *J. Clin. Invest.,* 68, 1125, 1981.
77. **Beard, C. J., Lyndon, K., Newburger, P. E., Ezekowitz, A. B., Arceci, R., Miller, B., Proto, P., Ryan, T., Anast, C., and Simons, E. R.,** A neutrophil defect associated with malignant infantile osteo-petrosis, *J. Lab. Clin. Med.,* in press.
78. **Horne, W. C. and Simons, E. R.,** Effects of amiloride on the response of human platelets to bovine thrombin, *Thrombosis Res.,* 13, 599, 1978.
79. **Pressman, B. C.,** Biological applications of ionophores, *Annu. Rev. Biochem.,* 45, 501, 1976.
80. **Thomas, J. A., Buchsbaum, R. N., Zimniak, A., and Racker, E.,** Intracellular pH measurements in Ehrlich ascites tumor cells utilizing spectroscopic probes generated in situ, *Biochemistry,* 18, 2210, 1979.
81. **Simchowitz, L.,** Chemotactic factor-induced activation of Na^+/H^+ exchange in human neutrophils. II. Intracellular pH changes, *J. Biol. Chem.,* 260, 13248, 1985.
82. **Simons, E. R., Rattoballi, R., and Proto, P.,** Flow cytometric measurements of intracellular pH in human platelets, *Fed. Proc.,* 45(abstr.), 852, 1986.
83. **Rink, T. J., Tsien, R. Y., and Pozzan, T.,** Cytoplasmic pH and free Mg^{++} in lymphocytes, *J. Cell Biol.,* 95, 189, 1982.
84. **Musgrove, E., Rugg, C., and Hedley, D.,** Flow cytometric measurement of cytoplasmic pH: a critical evaluation of available fluorochromes, *Cytometry,* 7, 347, 1986.
85. **Niedel, J.,** Detergent solubilization of the formyl peptide chemotactic receptor, *J. Biol. Chem.,* 256, 9295, 1981.
86. **Painter, R. G., Sklar, L. A., Jesaitis, A. J., Schmitt, M., and Cochrane, C. G.,** Activation of neutrophils by N-formyl chemotactic peptides, *Fed. Proc.,* 43, 2737, 1984.
87. **Niedel, J., Wilkinson, S., and Cuatrecasas, P.,** Receptor mediated uptake and degradation of ^{125}I-chemotactic peptide by human neutrophils, *J. Biol. Chem.,* 254, 10700, 1979.
88. **Scribner, D. J. and Fabirvey, D.,** Neutrophil receptors for IgG and compliment, *J. Immunol.,* 116, 892, 1976.
89. **Fearon, D. T.,** Identification of the membrane glycoprotein that is the C3b receptor of the human erythrocyte, polymorphonucleare leukocyte, B lymphocyte and monocyte, *J. Exp. Med.,* 152, 20, 1980.
90. **Fearon, D. T.,** Structure and function of the human C3b receptor, *Fed. Proc.,* 43, 2553, 1984.
91. **Boyum, A.,** Isolation of mononuclear cells and granulocytes from human blood, isolation of mononuclear cells by one centrifugation, and of granulocytes by combining centrifugation and sedimentation at 1 g, *Scand. J. Clin. Lab. Invest.,* 21, 97, 77, 1968.
92. **Holmsen, H.,** Secretable storage pools in platelets, *Annu. Rev. Med.,* 30, 119, 1979.

Chapter 8

PHOSPHOINOSITIDE METABOLISM

Holm Holmsen

TABLE OF CONTENTS

I. INTRODUCTION*

The numbers of reports on PPI metabolism during stimulation of secretion and other cellular events have been escalating rapidly over the last few years. This metabolism is quantitatively and qualitatively different from PPI metabolism in the unstimulated cell (Chapter 5) and appears to be tightly coupled to the agonist-induced mobilization of cytoplasmic Ca^{2+} and formation of DG, which in turn activate Ca^{2+}-dependent protein kinases and protein kinase C, respectively (Volume II, Chapter 10). In the first part of this review the PPI metabolism in stimulated cells in general will be discussed and particularly, the energy consumption involved will be addressed. In the second part the main enzymatic steps involved and their regulation will be discussed. The third part deals with the information available on PPI metabolism during stimulation of secretion in macrophages, neutrophils and mast cells, synaptosomes and other brain cells, various adrenal cell types, platelets, exocrine pancreas, and salivary glands. Finally, a brief evaluation of the methodological problems encountered in the study of PPI metabolism in these cells and some comments on future trends are presented.

II. PHOSPHOINOSITIDE METABOLISM IN STIMULATED CELLS

A. The PPI Cycle

The PPIs include the acidic phospholipids PI, PIP, and PIP_2 which account for about 4 to 10, 0.5, and 0.4%, respectively, of the phospholipids in a cell. All three inositol lipids seem to be present in most membrane systems of the cell, but since it is becoming increasingly clear that inositol lipid metabolism is compartmentalized (see below), it is possible that the function of PPI is different in each of these systems. Currently, it is believed that PI, PIP, and PIP_2 in the plasma membrane are metabolically interconverted in a cyclic fashion during cell stimulation as shown in Figure 1 based on recent reviews on the role of PPI metabolism in cellular signal processing in general.[1-6] According to this scheme (thick arrows), stimulation of the cell causes an instantaneous splitting of PIP_2 by a *phosphodiesterase* which yields DG and IP_3. It is believed that the PIP_2 consumed is rapidly replenished by two consecutive, ATP-requiring phosphorylations, the first of PI with *PI kinase* yielding PIP, and the second of PIP with *PIP kinase* yielding PIP_2. The DG, which accumulates transiently, is rapidly phosphorylated by the ATP-requiring *DG kinase* to PA. PA usually accumulates to relatively high concentrations, and is slowly converted back to PI by two reactions, the *PA:cytidylate transferase* reaction in which PA is converted to CDPDG by splitting of CTP to the transferred cytidylate and PP_i, and the *CDPDG:(myo)inositol transferase* reaction in which free inositol is added to the PA moiety of CDPDG with the release of CMP.

DG is also believed to be formed by phosphodiesteratic cleavage of PI and PIP, yielding IP and IP_2, respectively (Figure 1, thin arrows). In order to operate as a closed circuit (when one regards that ATP and CTP are derived from their respective pools in the cell), the PPI cycle has to provide free inositol. This is accomplished by three successive dephosphorylations of IP_3 to free inositol via IP_2 and IP (Figure 1, broken lines), reactions that will not be discussed in this review. It is only over the last 5 to 6 years that the full cyclic metabolism of the PPIs, as illustrated in Figure 1, has been recognized. Earlier studies on signal transduction in cells using Ca^{2+} as second messenger were thought to involve the "PI cycle",

---

*   Abbreviations: ATP = adenosine 5'-triphosphate; CDPDG = cytidine 5'-diphosphate diacylglycerol; CTP = cytidine 5'-triphosphate; DG = diacylglycerol; GTP = guanosine 5'-triphosphate; IP = inositol 1-phosphate; IP_2 = inositol 1,4-bisphosphate; IP_3 = inositol 1,4,5-trisphosphate; PA = phosphatidic acid; PI = phosphatidylinositol; PIP = phosphatidylinositol 4'-phosphate; PIP_2 = phosphatidylinositol 4',5'-bisphosphate; P_i = inorganic orthophosphate; PPI = polyphosphoinositide; PP_i = inorganic pyrophosphate;

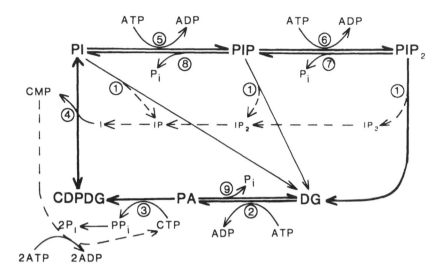

FIGURE 1. PPI metabolism in stimulated cells — the PPI cycle. The number of the individual enzymes are encircled to show the exact reaction they are catalyzing. The details of the cyclic metabolism (thick arrows), side reactions, and alternative reactions (thin arrows) and the hydrolytic conversion of IP_3 (broken lines) to free inositol (I) are given in the text.

or the "PI response", which includes the cyclic pathway from PI via DG, PA, CDPDG, and back to PI. However, no convincing evidence for the formation of a Ca^{2+}-liberating substance or reaction was ever found with the PI cycle.

The PPI cycle (Figure 1) generates the unique substance IP_3 which is said to possess Ca^{2+}-mobilizing properties (for reviews, see References 3 to 6). The release of Ca^{2+} from nonmitochondrial sources by IP_3 has been demonstrated in a number of permeabilized secretory cells such as pancreatic acinar cells,[7] insulin-secreting RINm5F cells,[8] macrophages,[9] mammotropic pituitary cells,[10] and platelets.[11-15]

B. Energy Aspects of the PPI Cycle

The resynthesis of PIP and PIP_2 are rapid processes, and the concentrations of PIP and PIP_2 undergo initial, transient decreases upon stimulation and thereafter tend to increase above prestimulation levels. The resynthesis of PI from PA, however, is relatively slow so that the most marked change in the steady-state levels of the metabolites in the PPI cycle after cell stimulation is a decrease in PI and an increase in PA. This is illustrated in Figure 2 by the time course of changes in PPI metabolites after stimulation of platelets with thrombin. However, these time courses do not give any information about the mechanism of PI disappearance: is it converted to DG by direct phosphodiesteratic cleavage or does it disappear by phosphorylation(s) to PIP_2 which is being constantly converted to DG and IP_3? Thrombin-induced PI disappearance in platelets is abolished by ATP depletion[16,17] which favors the phosphorylation route. Although we lack information about the flux through the PPI cycle during cell stimulation, Figures 1 and 2 suggest that for each PIP_2 molecule cleaved, three molecules of ATP are consumed: two for resynthesis of PIP_2 from PI and one for PA formation. For one whole turn of the PPI cycle (PI to PI) two additional ATP molecules are consumed to replenish the CTP used in the formation of CDPDG. Thus, five molecules of ATP are consumed in one turn of the PPI cycle. In platelets it has been estimated that PI turnover increases from 3.2 to 21 nmol/min/10^{11} cells upon maximal thrombin stimulation.[18] Thus, if this increase in PI turnover of stimulated platelets is caused by complete cycling of the PPI cycle illustrated in Figure 1, stimulation of platelet secretion must be accompanied by an ATP consumption of about 100 nmol/min/10^{11} cells.

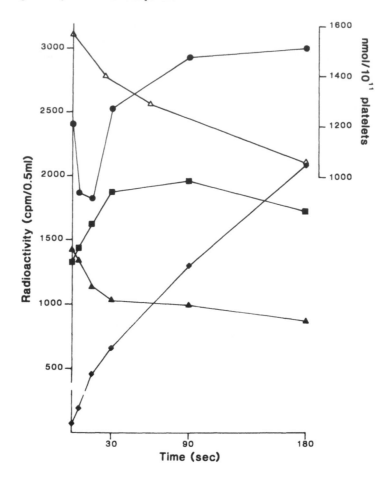

FIGURE 2.   Changes in the levels of ^{32}P-labeled phosphoinositides in thrombin-stimulated platelets. Human platelets were labeled in (citrated) platelet-rich plasma with 0.3 mCi/mℓ of ^{32}P-orthophosphate for 1 hr (37°C) and gel-filtered into Tyrode's solution containing 0.2% albumin and 5 mM glucose, but no CaCl$_2$ and phosphate. After the gel-filtered platelets had been incubated at 37°C for 90 min, 0.2 U/mℓ of thrombin was added (zero time on the abscissa) and samples were taken at the times indicated and analyzed for the radioactivity of PIP$_2$ (●), PIP (■), PI (▲), and PA (◆) as well as the mass (measured as phosphorus) of PI (△). The figure is based on results presented in Reference 18.

There are, however, three "futile" (sub)cycles in the PPI cycle in which one phosphoryl group of ATP is consumed in the formation of a phosphate ester by a kinase while at the same time this ester is hydrolyzed by a phosphomonoesterase to yield P$_i$; the net result of the kinase/phosphohydrolase reactions is splitting of ATP to ADP + P$_i$. The enzyme pairs constituting such ATP-consuming reactions in the PPI cycle are PI kinase/PIP$_2$ phosphomonoesterase, PIP kinase/PIP$_2$ phosphomonoesterase, and DG kinase/PA phosphomonoesterase (Figure 1). Very little is known about the consumption of ATP in these kinase/hydrolase reactions. From Chapter 5 we learn that maintenance of PIP and PIP$_2$ in the phosphorylated state (i.e. phosphorylation and dephosphorylation of the monophosphoesters of the PPIs) requires 7 and 1% of the total ATP consumption in resting platelets and erythrocytes, respectively. Whether this ATP consumption increases or decreases during stimulation is unknown.

Thus, although increased PPI metabolism clearly represents part of the total energy requirement during stimulation of secretion, we are far from knowing the exact magnitude of the energy consumption. The above estimates for platelets were made on the assumption that turnover of PI represents turnover of the entire PPI cycle. Results in our laboratory suggest that the bulk of the PI pool in platelets does not participate in the reactions shown in Figure 1. If this proves to be the case, the measured turnover in PI[18] is probably caused by the "PI cycle" which consumes three molecules of ATP per turn, and the increased turnover due to stimulation would amount to about 60 nmol ATP/min/10^{11} platelets.

III. THE ENZYMES IN THE PPI CYCLE

Very little is known about the individual enzymes of the PPI cycle in secretory cells. The greater part of our rather limited knowledge about these enzymes and their regulation comes from studies of liver, brain, and red cells that are unrelated to secretion. In the following pages the literature is reviewed with reference to the particular cell type used. In several of the studies cited the amino glycoside antibiotic neomycin has been used as a "specific" inhibitor for reactions with PPIs as substrate. This inhibition is based on the high affinity of neomycin to PIP and particularly PIP_2.[19-22] The specificity of neomycin in cellular systems has recently been questioned since it strongly binds (free) ATP.[23] However, $MgATP^{2-}$, and not free ATP^{4-}, is the species that is present intracellularly; it is not known if neomycin is capable of binding $MgATP^{2-}$ and rendering cellular ATP unavailable.

A. Phosphodiesterase

The phosphodiesterase, also called PI- or PPI-specific phospholipase C, hydrolyzes PI, PIP, or PIP_2 phosphodiesteratically to DG and, respectively, to IP, IP_2, or IP_3 and is enzyme no. 1 in Figure 1. The best characterized enzyme has been purified from the soluble phase of liver,[24] seminal vesicles,[25] brain,[26] and platelets[27] and assayed with PI as substrate. The enzyme is heterogeneous with mol wt ranging from 68,000 to 140,000, is active at neutral pH, and is dependent on both Ca^{2+} and phospholipid. Whether these enzymes also cleave PIP and PIP_2 is not yet clear. The PI which is hydrolyzed diesteratically during stimulation of neutrophils has been localized to the plasma membrane of these cells;[28] however, a cytosolic localization of the phosphodiesterase cannot be excluded.

The two PI phosphodiesterases purified from sheep seminal vesicles were more active with PIP and PIP_2 as substrates than PI; the three substrates were competitive with PI having the highest affinity; most interestingly, the enzymes hydrolyzed PIP and PIP_2 in the presence of EGTA, but not PI.[29] Except for this study with purified enzyme, PPI phosphodiesterase(s) has until now been studied as the disappearance of the endogenous PIP or PIP_2 from membranes upon Ca^{2+} addition. Diesteratic PIP_2 hydrolysis was observed in the plasma membrane from human granulocytes incubated with the secretagogue fMet-Leu-Phe in the presence of Ca^{2+}; this hydrolysis was markedly stimulated by GTP and the stimulation was blocked by pertussis toxin.[30] A role for GTP, presumably through a G-protein, in the activation of PPI phosphodiesterase has also been suggested from experiments with permeabilized platelets,[31-33] isolated neutrophil plasma membranes,[33] and rat hepatocytes.[35] It is therefore surprising that PIP_2-specific phosphodiesterase is associated with soluble, granule-binding proteins in the adrenal medulla,[36] a tissue where secretion is thought *not* to be mediated through PPI metabolism (see below).

An enzyme present in human, rabbit, and rat erythrocyte ghosts released IP_2 and IP_3 from endogenous PIP and PIP_2, respectively, when the ghosts were incubated with Ca^{2+}; the activity which was inhibited by neomycin and unaffected by Mg^{2+} could not be demonstrated in ghosts from pig and ox erythrocytes.[37] Neomycin also caused a parallel inhibition of GTP-stimulated PIP_2 and histamine secretion from permeabilized mast cells.[34] A PIP_2 phosphodiesterase present in the soluble fraction of rat brain cells exhibited an absolute dependence for Ca^{2+} with maximal activity at $10^{-7}\ M\ Ca^{2+}$ when assayed with pure PIP_2; however, when PIP_2 was presented as a component of the inner leaflet of a biological membrane in a solution with the same ionic environment as cytoplasm, maximal activity was obtained at 10^{-3} M Ca^{2+} and no activity was detected with 10^{-4} M.[38] It was further shown that the enzymatic activity depended markedly on the configuration of the lipid micelles containing the substrate PIP_2. PIP_2 phosphodiesterase from artery smooth muscle was also activated by Ca^{2+} and markedly inhibited by Mg^{2+}; this inhibition was counteracted by GTP.[39] A membrane fraction obtained from nerve endings has been shown to contain bound, Ca^{2+}-stimulated PPI phosphodiesterase.[40]

B. DG Kinase

DG kinase catalyzes transfer of one phosphoryl group from ATP to DG by which PA is formed. The enzyme is no. 2 in Figure 1, and the step consumes energy directly.

The kinase has been well characterized in *Escherichia coli* where it is solely bound to the membrane; its primary structure has been established.[41] The kinase has been demonstrated in a variety of mammalian tissues, both in a soluble and membrane-bound form, and displayed little specificity with regard to the fatty acid composition of substrate DG.[42,43] The cytosolic enzyme from rat liver has been purified 2000-fold, has an apparent mol wt of 121,000 and K_m = 170, 540, and 3 × 10^{-6} M for ATP, GTP, and DG, respectively.[42] This enzyme showed an absolute requirement for Mg^{2+} and detergent (deoxycholate) and was inhibited by ADP. The soluble enzyme from pig brain has also been purified and had a mol wt of 77,000; this enzyme required Mg^{2+} and phospholipid or detergent for activity.[43] PA which was formed in stimulated neutrophils accumulated in the plasma membrane,[28] indicating the presence of DG kinase in this membrane.

It has recently been found that the transforming gene product from Rous sarcoma virus ($pp60^{v-src}$) has DG kinase activity which may be regulated in the same way as the casein kinase activity of this protein.[44] Epidermal growth factor activated the DG kinase, but not PIP kinase of isolated plasma membranes tested with exogenous DG with or with detergent; the activation was absolutely dependent on Mg^{2+}.[45]

C. Phosphatidate Phosphohydrolase

PA phosphohydrolase catalyzes the hydrolytic cleavage of PA to DG and P_i, and is reaction no. 9 in Figure 1. Together with the DG kinase reaction this hydrolysis represents one of the futile cycles in PPI metabolism where ATP is hydrolyzed in an apparently purposeless manner. PA phosphohydrolase appears to exist in Mg^{2+}-dependent and -independent forms in various subcellular fractions of most cells investigated. Both forms in platelets[46] are inhibited by trifluoperazine, apparently by association of the phenothiazine with substrate PA.[47] The enzyme in rat adipose tissue is dependent on Mg^{2+} and activated by the polyamines spermine, spermidine, and putrescine.[48]

D. CTP:Phosphatidate Cytidyl Transferase

This enzyme catalyzes the transfer of the CMP moiety from CTP to PA with the formation of CDPDG (also referred to as CMP-PA) and PP_i, and represents reaction no. 3 in Figure 1. It has been studied in at least one secretory cell, the human platelet,[49] and found to be similar to the enzyme in rat liver.[50] The transferase requires Mg^{2+}, is tightly associated with membrane structures, and is specific for CTP and PA, although it displays little selectivity with regard to the fatty acid composition of the latter. The transferase reaction is highly reversible in principle, but in the cell the reaction is driven towards CDPDG formation as PP_i is rapidly removed by inorganic pyrophosphatase (Figure 1).

E. CDPDG:Inositol Phosphatidyl Transferase

This enzyme transfers the PA moiety of CDPDG to (free) myoinositol with the formation of PI and CMP; the reaction is no. 4 in Figure 1. In the PPI cycle free myoinositol is provided by the "closed circuit" shown by the broken lines in the middle of Figure 1. The enzyme has apparently not been studied in any of the secretory cells discussed in this review, and little is known about its characteristics beyond its original description in brain synaptosomes almost 30 years ago.[51] As discussed briefly below, most of our knowledge of PPI metabolism stems from studies with radioisotopic prelabeling of cells, and the use of ^{3}H-inositol has revealed that many types of cells incorporate the inositol label into PI independent of CDPDG via the "PI:inositol exchange enzyme" both in the absence and presence of Mn^{2+}.[52,53] Most importantly, Mn^{2+} causes the incorporation of inositol into a pool of PI that is not coupled to PPI (or PI) hydrolysis.[52,53]

F. PI Kinase

This enzyme catalyzes reaction no. 5 in Figure 1 in which ATP is directly consumed by transfer of its γ-phosphoryl to the 4 position on the inositol ring of PI with the formation of ADP.

PI kinase has been demonstrated in both the plasma membrane and the granule membrane of human granulocytes.[30] The enzyme activity has also been found in plasma membranes from pig granulocytes and shown to be particularly stimulated by the catalytic subunit of cAMP-dependent protein kinase in the presence of ATP. The stimulation is associated with the phosphorylation of a 24,000 M_r protein.[54] Plasma membranes of lymphocytes, platelets, and red cells also contain a PI kinase which is similarly stimulated.[55,56] This enzyme has been demonstrated in purified preparations of microsomes,[57] Golgi,[58,59] lysosomal membranes,[59,60] and nuclear envelopes.[61] PI kinase which shows absolute dependence on Mg^{2+} and ATP, and which phosphorylates exogenously presented PI has been shown to be tightly associated with coated vesicles.[62]

The Rous sarcoma virus transforming gene product has PI (and DG) kinase activity.[44] Purified insulin receptor preparations from human placenta also possess PI kinase activity.[63] These activities are inhibited in the same way as the associated tyrosine kinase activity.[44,63] The same association of PI kinase and tyrosine kinase was found for the epidermal growth factor receptor, but these kinase activities could be separated as two protein entities.[64]

G. PIP Kinase

PIP kinase catalyzes reaction no. 6 in Figure 1 in which ATP is directly consumed by transfer of its γ-phosphoryl to the 5 position on the inositol of PIP with the formation of ADP.

The PIP kinase demonstrated in erythrocyte ghosts was found to have a distinctly higher affinity for Mg^{2+} than the PI kinase in these membranes.[65] An Mg^{2+}-dependent enzyme in the high speed supernatant of bovine retina was powerfully inhibited by its own product, PIP_2. Phosphodiesterase was also present in this extract, which, when activated by Ca^{2+}, caused the relief of the PIP_2 inhibition (release of feedback inhibition results in replenishment of PIP_2).[66] PIP kinase was present in plasma membranes, but not in granule membranes from human[30] and pig granulocytes.[54] PIP kinase has also been demonstrated in nuclear envelopes.[67] In rat liver, PIP kinase was found to be predominantly localized in the plasma membrane[50,68] In contrast to PI kinase which was found in the Golgi, lysosomal membranes, and plasma membranes.[59]

PIP kinase has been purified 67-fold from rat brain (both cytosol and plasma membranes) and was characterized as a protein with mol wt of approximately 45,000; it had an isoelectric point of 5.8 and was activated by Mg^{2+}.[69] Antibody raised against this protein inhibited the PIP kinase activity.[70] A 48-kdalton protein ("B-50", isoelectric point 4.5) from rat brain which is phosphorylated by protein kinase C has been shown to inhibit semipurified PIP kinase, while the dephosphorylated form had little effect.[71]

H. PIP and PIP₂ Phosphomonoesterases

These enzymes, which may be different enzyme proteins, catalyze the hydrolytic cleavage of the 4'-monoesterphosphate of PIP (PIP phosphomonoesterase) and the 5'-monoesterphosphate of PIP₂ (PIP₂ phosphomonoesterase); they are represented by reactions 8 and 7, respectively, in Figure 1.

Nuclear membranes readily degrade endogenous PIP or added PIP (after solubilization with Tween 80) at neutral pH with a corresponding formation of inorganic orthophosphate in the presence of EDTA; Mg^{2+}, Ca^{2+}, polyamines, neomycin, and inositol 2-phosphate were found to be inhibitory, while known inhibitors of acid or alkaline phosphatase or the substrates of these enzymes had no effect.[72] PIP phosphomonoesterase from kidney[73] and erythrocyte[74] also do not require divalent cations, which is in contrast to the membrane-bound enzyme studied in rat kidney cortex[75] which requires Mg^{2+}, and the microsome-bound enzyme in rabbit iris muscle[76] and rat brain[79] or erythrocyte cytosol[78] which were inhibited by EDTA and activated by Mg^{2+} and Ca^{2+}. PIP₂ phosphomonoesterase from rat brain (soluble phase) hydrolyzes PIP₂ presented in the biological membrane and suspended in a medium having the ionic composition of cytoplasm in a Mg^{2+}-dependent manner; Ca^{2+} did not influence this activity.[38] An Mg^{2+}-dependent phosphomonoesterase specific for PIP₂ has been described in erythrocyte membranes.[79]

A phosphomonohydrolase from erythrocyte cytosol has been purified 27,000-fold and shown to be specific for PIP₂; it has a mol wt of 105,000, required Mg^{2+}, was relatively unaffected by SH-reacting agents, and was inhibited by Ca^{2+} and 10^{-6} to 10^{-4} M neomycin in the absence of nonionic detergent.[78] In contrast, erythrocyte membranes contain a specific phosphomonohydrolase for PIP which is fully active in the presence of EDTA and sensitive to SH-reacting agents.[74] Such membranes have also been shown to contain a Ca^{2+}-requiring phosphomonoesterase which is inhibited by Mg^{2+}.[80] A cytosolic monoesterase in erythrocytes which is activated by Ca^{2+} has also been described.[81]

IV. PPI METABOLISM IN SPECIFIC SECRETORY CELLS

In the following sections, experimental evidence will be cited that seems to indicate the involvement of steps in the PPI metabolism in initiation of secretory mechanisms in specific secretory tissues and cells. However, the evidence for coupling of PPI metabolism to either receptor occupancy or secretion is not discussed.

A. Macrophages, Neutrophils, and Mast Cells

1. Macrophages

Macrophages secrete lysosomal enzymes when stimulated with formylated chemotactic peptides such as fMET-Norleu-Phe or fMet-Leu-Phe. Stimulation of secretion with peptide caused a decrease in the level of radioactive PI, PIP, and PIP₂ and formation of radioactive PA (and DG in the case of ³H-glycerol) in macrophages prelabeled with ³²-P-orthophosphate or ³H-glycerol.[82] Transient breakdown of ³²P-PIP₂ and -PIP together with a massive accumulation of ³²P-PA which was independent of cytoplasmic Ca^{2+} levels, has been demonstrated in ³²P-phosphate-labeled peritoneal macrophages stimulated with peptide.[83]

2. Neutrophils

Neutrophils secrete lysosomal enzymes in parallel with an increase in cytosolic Ca^{2+} when stimulated with chemotactic peptides such as fMet-Leu-Phe.[84] Both breakdown and resynthesis of PI and accumulation of PA take place during secretion by rabbit[85,86] and human[87] neutrophils, although these changes were thought to be a consequence of influx of extracellular Ca^{2+} rather than the cause of this influx. With human neutrophils no effect of prevention of the Ca^{2+} influx on peptide-stimulated PI breakdown and PA formation was found.[88] Also with guinea pig neutrophils, Ca^{2+} deprivation had no effect on peptide-induced incorporation of ^{32}P into PI and PA.[89]

Both the peptide and pseudomonal leukocidin caused a rapid formation of PA and a transient breakdown of PIP and PIP_2 in rabbit neutrophils that appeared to correlate with Ca^{2+} movements.[90-93] Similarly, secretion of enzyme induced by fMet-Leu-Phe in human normal and leukemic ^{32}P-labeled cells was associated with a rapid disappearance of radioactive PI and PIP_2 and accumulation of radioactive PA and PIP.[94]

3. Mast Cells

Mast cell secretion of histamine is dependent on extracellular Ca^{2+} when stimulation is by IgE-directed ligands, the divalent cationophore A23187, and compound 48/80 which apparently acts at the receptor level. It was found that although the two latter compounds induced similar influxes of Ca^{2+} and secretion, A23187 caused incorporation of radioactivity in DG, PA, and PI in mast cells prelabeled with ^3H-glycerol while compound 48/80 did not.[95] These findings were interpreted as evidence for a coupling of *de novo* synthesis of PI to secretion; however, the changes in isotope labeling lagged distinctly behind secretion, and could as well be a consequence of secretion when specifically stimulated with A23187. With IgE antibody-induced secretion, a very rapid, transient decrease in the radioactivity of PIP and later in PI (but not in PIP_2) was found in cells prelabeled with ^3H-glycerol or arachidonate; only a (late) increase in PI took place in cells prelabeled with ^{32}P-orthophosphate.[96] A parallel inhibition of PPI hydrolysis and secretion by pertussis toxin[97] further showed that increased PPI metabolism represents energy-consuming processes during secretion in mast cells. Neomycin completely inhibited histamine secretion when introduced into mast cell cytosol.[34]

B. Synaptosomes and Other Brain Cells

Incorporation of ^{32}P-orthophosphate into PI was demonstrated upon carbamylcholine stimulation of neuroblastoma cells and this PI effect was abolished by atropine.[98] Neuronal depolarization caused Ca^{2+} uptake and a concomitant PI-PA conversion, both of which were inhibited by phospholipase C inhibitors; extracellularly added PA also caused Ca^{2+} uptake that was unaffected by these inhibitors.[99] Increased PI metabolism has been observed upon occupation of the 5HT 2, but not the 5HT-1 receptors in cerebral cortex.[100] IP_3 accumulated rapidly in isolated rat superior cervical sympathetic ganglia when V_1-vasopressin receptors were occupied and during muscarinic cholinergic stimuli[101] and depolarization[102] which suggested that hydrolysis of PIP_2 was occurring.

Inositol 1,3,4,5-tetrakisphosphate has recently been reported to be formed during stimulation of cerebral cortical slices, but the origin and significance of this inositol polyphosphate is not known; however, it could be a precursor for IP_3.[103]

C. Adrenal Cells
1. Medulla

The secretion of catecholamines, opioid peptides, dopamine β-hydroxylase, and chromogranins from the chromaffin cells of the adrenal medulla has recently been reviewed.[104] The major agonist acetylcholine, which is released from stimulated splanchnic nerve endings, acts via nicotinic receptors to cause opening of voltage-dependent Ca^{2+} channels apparently independent of PPI metabolism;[105] the internalized Ca^{2+} is then thought to induce exocytosis through actin dissociation, calmodulin binding to the granule surface, and protein phosphorylation, although this central role of Ca^{2+} is not entirely clear. Acetylcholine also has an inhibitory effect on exocytosis which is mediated via muscarinic receptors with formation of cGMP and diesteratic hydrolysis of PI.[106] Inhibitory responses by muscarinic receptors coupled to PPI hydrolysis may be a general phenomenon, possibly mediated by DG-induced activation of protein kinase C.[107] Stimulation of muscarinic receptors in chromaffin cells was also shown to be accompanied by accumulation of PA, efflux of Ca^{2+}, and synthesis of cGMP; the same effect was obtained by addition of PA to the cells, indicating that PA also could be responsible for inhibition of exocytosis.[108] Fragmentary evidence for an involvement of PPI hydrolysis in exocytosis by adrenal medullary cells has been published.[109]

2. Glomerulosa Cells

Aldosterone secretion by the adrenal glomerulosa cells is stimulated by both ACTH and angiotensin II, but only secretion by the latter agonist is accompanied by increased PI metabolism.[110-112] Li^+ ions, which block dephosphorylation of IP,[113] completely prevented both secretion and PI resynthesis (measured as ^{32}P incorporation) indicating that replenishment of PI in the cyclic metabolism (Figure 1) is necessary for the stimulus-secretion coupling.[114] A rapid and extensive decrease in the level of ^{32}P- and ^3H-inositol-labeled PIP_2 was found in rat adrenal glomerulosa cells stimulated with the Ca^{2+}-mobilizing angiotensin II, but not with the cAMP-mobilizing ACTH.[115]

D. Platelets

Platelets secrete the contents from three different storage granules when stimulated with agents (e.g. thrombin, collagen, ADP) that also cause platelet shape change, aggregation, and liberation of arachidonate esterified in phospholipids. One type of storage granules is the dense granules which contain ATP, ADP, serotonin, and Ca^{2+} (human platelets); the second type is the α-granules which contain proteins such as fibrinogen, factor V, Von Willebrand factor, platelet-derived growth factor, and the specific platelet proteins β-thromboglobulin and platelet factor 4; the third type of storage granules contains lysosomal enzymes, particularly acid glycosidases. The induction of the secretory responses by thrombin has been extensively studied and shown to be associated with the PPI metabolism shown in Figures 1 and 2.[116,117] Recently, occupation of the serotonin-S_2 receptor has been reported to be similarly coupled to the PPI metabolism in platelets.[118]

Platelet secretion is powerfully inhibited by agents that elevate cAMP levels in platelets and this inhibition is paralleled by inhibition of PI disappearance and PA accumulation,[119] metabolism of PIP and PIP_2,[120] and formation of inositol phosphates.[121] This inhibition is difficult to reconcile with the findings that the catalytic subunit of cAMP-stimulated protein kinase powerfully stimulates PI kinase in platelet membranes.[56]

The formation of inositol phosphates in platelets during stimulation has been demonstrated by measuring ^3H-labeled phosphates after preincubation with ^3H-inositol;[11-15,122] in one study ^3H-inositol 4-phosphate was found[123] suggesting preferential hydrolysis of the 1- and 5-phosphate of IP_3 or IP_2, or the existence of PI with a 4'-phosphate bond. By analysis of mass, it was found that the amounts of IP_3 formed were substantially smaller than the DG formed simultaneously which suggests that PI, in particular, and perhaps also PIP are hydrolyzed.[124] Stimulation of platelets in the presence of excess ^{32}P-P_i gave the same increase in the specific radioactivity in PIP_2 as in PA,[125] which is in equilibrium with the γ-phosphate of ATP,[17] suggesting that little resynthesis of PIP_2 takes place, as the PPI cycle (Figure 1) predicts. The equal increase in the specific radioactivity of ^{32}P in PIP_2, PA, and ATP has recently been found to be caused by a massive increase in net phosphate influx from the medium immediately upon platelet stimulation.[126]

Thus, from the studies with platelets (and pancreas below) evidence is now appearing which indicates that the PPI cycle, as shown in Figure 1, may be oversimplified. If considerably more DG is formed than can be accounted for by IP_3 formation, and if PIP_2 resynthesis is minimal, the marked decrease in PI mass (Figure 2, Reference 117) upon stimulation is probably due to direct diesteratic hydrolysis of PI and not, as indicated in Figure 1, due to successive phosphorylation only. The PA formed during platelet stimulation is metabolically heterogeneous as ^3H-PA is made distinctly later than ^{32}P-PA in platelets prelabeled with both ^{32}P-P$_i$ and ^{3}H-arachidonate.[118,127] This metabolic heterogeneity could be due to different pools of DG, one originating rapidly from PIP$_2$ and the other from PI. However, irrespective of origin, DG would be converted to PA by DG kinase. By the use of R59 022, an inhibitor of DG kinase, only the initial formation of PA was blocked, indicating that the formation of PA in the later period was not via DG kinase.[128] The late fraction of PA could be formed by *de novo* synthesis (see below).

Phorbol ester (TPA) causes changes in PIP and PIP$_2$ in platelets[129,130] that are very similar to those produced by thrombin (Figure 2) except that PA is not formed and the cells show no responses during the rapid PPI metabolism.[129] DG produces the same effect on PPI metabolism as TPA.[131] Since no PA is formed it is reasonable to assume that the TPA- and DG-induced changes were caused by a shift in the relative activities of the kinases and phosphohydrolases that regulate the steady-state levels of PIP and PIP$_2$. Since DG is formed during thrombin stimulation of platelets, it is possible that it affects PIP/PIP$_2$ homeostasis in the same way as exogenous DG (or TPA) in addition to the effects caused by (thrombin-induced) stimulation of the phosphodiesterase. TPA[124,132] and DG[132] have been reported to inhibit phosphodiesterase activation by thrombin as well as PA formation and Ca^{2+} mobilization.[133] Whether this inhibition is caused by their effect on PPI metabolism or via activation of protein kinase C, or both, is not resolved.

Chlorpromazine and other cationic amphophilic drugs also have pronounced effects on PPI metabolism. In resting platelets pulse-labeled with ^{32}P-P$_i$ the drugs cause an increase in the ^{32}P-content of PIP[134,135] and chlorpromazine enhances the thrombin-induced elevation of ^{32}P-PIP and inhibits secretion.[135]

E. Pancreas

The secretion of zymogens from the exocrine pancreas was the secretory process in which increased PI metabolism was first discovered. In 1953 Hokin and Hokin[136] showed that phospholipids were strongly labeled when secretion was conducted in the presence of ^{32}P-orthophosphate and demonstrated later that PA and PI were the specific phospholipids to be labeled.[137] They first thought that this increased metabolism was coupled specifically to the exocytotic event, but studies by these and other investigators with pancreas and other cells, as reviewed by Michell[138] and Laychock and Putney,[2] suggested that the increased PI metabolism represented a biochemical mechanism that coupled receptor activation to Ca^{2+} mobilization. Disappearance of PI during carbachol-stimulated secretion has been demonstrated and thought to be caused both by phospholipase C and phospholipase D.[139] Rapid decrease in PIP$_2$ was found to be an initial event during stimulation with carbachol or pancreozymin and this decrease was accompanied by accumulation of PA[140] and inositol trisphophate.[141] Farese et al.[142] have suggested that hydrolysis of PI and PIP$_2$ are separable processes in the pancreas and only PIP$_2$ hydrolysis, which is obtained at high agonist concentrations and accompanied by^{32} P incorporation into PI and PA, is associated with receptor-controlled Ca^{2+} mobilization. Further evidence for direct phosphodiesteratic cleavage of PI comes from the demonstration of formation of inositol 1,2-cyclic phosphate in this tissue.[143]

The pancreatic islets, from which the endocrine secretion of insulin and glucagon occurs, also incorporated ^{32}P into PI and PA during stimulation with carbamylcholine in islets that had been prelabeled with ^{32}P-orthophosphate.[144] In cells prelabeled with ^{3}H-inositol, a decrease in ^{3}H-phospholipid and an increase in ^{3}H-inositolphosphates was demonstrated following stimulation with glucose and carbamylcholine.[145] The same authors also demonstrated increased incorporation of ^{3}H-glycerol into islet PA and PI, while incorporation into PC and PE decreased during stimulation of insulin secretion; they concluded that *de novo* synthesis of PI and PA took place in addition to increased PI cycle activity.[146]

In pancreatic acinar cells, as well as in hepatocytes and certain leukemic cells, stimulation with agonists that raised the level of cytoplasmic Ca^{2+} caused formation of the inositol 1,3,4-trisphosphate isomer in greater proportions than the "common" 1,4,5-trisphosphate (IP_3) isomer; the function of the former trisphosphate is not known.[147] The same isomer accumulated during stimulation of salivary glands.

F. Salivary Glands

The parotid gland is an exocrine secretory organ which responds to a variety of agonists with secretion of digestive enzymes and efflux of K^+. Early studies indicated that in the rat parotid incorporation of ^{32}P into PI[148] and PA[149] is associated with K^+ efflux via stimulation of α-adrenergic receptors. Peptidergic[149] and cholinergic[150] receptors in this gland are also coupled to PI turnover. Further studies have indicated that a closer relationship exists between PI metabolism and Ca^{2+} gating than receptor occupancy[151] and that the PA formed may be a primary mediator of this gating.[152] Breakdown of PIP_2, but not of PIP, was later reported to occur in rat parotid acinar cells in response to epinephrine (α-adrenergic), methacholine, and substance P in a manner that was independent of Ca^{2+} influx, indicating that PIP_2 breakdown is coupled to agonist-induced release of Ca^{2+} from intracellular sources.[153,154] Evidence for the formation of IP_3 and IP_2, but not IP, immediately upon methacholine stimulation has also been obtained.[155] However, a substantial proportion of the IP_3 formed was found to be the L -(myo)inositol 1,3,4-trisphosphate isomer, and not the D -(myo)inositol 1,4,5-isomer expected from hydrolysis of PIP_2; whether the formation of the 1,3,4-isomer originates by isomerization of the 1,4,5-isomer formed by phosphodiesteratic cleavage of PIP_2 or directly from a putative phosphatidylinositol 3,4-bisphosphate remains to be clarified.[156]

In rat submaxillary glands cholinergic stimulation caused a marked decrease in PI and an increase in PA masses in the presence of extracellular Ca^{2+}, but only changes in the ^{32}P content of these lipids in the absence of Ca^{2+}, suggesting two pools of hormone-sensitive PI and PA.[157]

In isolated blowfly salivary glands 5-hydroxytryptamine (serotonin) causes fluid secretion and transepithelial Ca^{2+} flux. Early studies showed that concomitant with these responses a massive breakdown of radioactive PI took place in glands briefly preincubated with ^3H-inositol or ^{32}P-orthophosphate, and that this breakdown was absent when fluid secretion was stimulated by A23187-mediated entry of Ca^{2+}; these observations strongly supported the concept that Ca^{2+} entry caused by serotonin was coupled with (radioactive) PI breakdown.[158] These glands became desensitized by prolonged incubation with serotonin in the presence of extracellular Ca^{2+}, a condition which prevented resynthesis of PI; the glands recovered during incubation with inositol in the absence of serotonin.[159] However, there was no measurable decrease in the PI mass during secretion, which showed that only a small pool of PI with rapid turnover was coupled to the gating of Ca^{2+}.[160] Further studies revealed that ^3H-DG accumulated transiently during stimulation of secretion in glands prelabeled with ^3H-arachidonate.[161] IP_3 and IP_2 accumulated distinctly much earlier than IP during serotonin stimulation of the glands, clearly demonstrating that PIP_2 and PIP were hydrolyzed before PI,[162] and such hydrolysis was shown to be independent of extracellular Ca^{2+}.[163] The desensitization described above was further shown to occur in parallel with a restoration of the PIP_2 consumed by conversion of PI to PIP_2.[164] Recently, the stimulation of phosphodiesteratic breakdown by 5-methyltryptamine has been shown to occur through a GTP-dependent mechanism in a cell-free system from blowfly salivary glands.[185]

G. Pituitary Glands

A number of peptide hormones are secreted from the pituitary gland by appropriate stimuli. The release of prolactin by TRH (thyrotropin-releasing hormone) and bombesin was accompanied by massive labeling of PA and PI in ^{32}P-labeled GH[166] and GH_3[167] cells. With cloned

GH₃ cells TRH caused disappearance of PI and a concerted formation of DG, PA, IP, and free inositol.[168] Inhibition of prolactin secretion by dopamine caused a parallel inhibition of ^{32}P-incorporation in PI which was shown not to be a consequence of mobilization of Ca^{2+} or cAMP.[169] In a gonadotroph-enriched culture of anterior pituitary cells, lutenizing hormone-releasing hormone and pancreatic growth hormone-releasing factor caused a rapid incorporation of ^{32}P into PA and subsequently into PI.[170]

A very rapid and extensive breakdown of ^{3}H-PIP₂ and ^{3}H-PIP took place in GH₃ pituitary tumor cells prelabeled with ^{3}H-inositol upon treatment with TRH: accumulation of DG and PA was seen in cells labeled with ^{3}H-glycerol and ^{32}P-orthophosphate, respectively, as well as ^{3}H-labeled IP₃, IP₂, IP, and inositol, which is typical for an initial phosphodiesteratic hydrolysis of the PPIs.[171,172] Similar changes in the ^{32}P-PPIs by TRH were reported for clonal GH₃ cells prelabeled with ^{32}P-Pᵢ.[173] Recently, an involvement of GTP in TRH-stimulated PPI hydrolysis has been demonstrated in isolated membranes from GH₃ cells.[174]

H. Others

Extracellular Ca^{2+} stimulated glucanase secretion from sea urchin eggs concomitant with PPI breakdown; both processes were inhibited by neomycin, and the PPI breakdown was thought to be coupled to the fusion step of the exocytotic event.[175]

In lacrimal acinar cells secretion is induced by muscarinic cholinergic agnoists, which also caused rapid PIP₂ breakdown, PA accumulation, and formation of inositol phosphates.[176]

V. CENTRAL PROBLEMS IN PPI METABOLISM RESEARCH

A. Metabolic Compartmentation

Figure 1 gives the impression that the *entire* pools of PI, PIP, and PIP₂ are participating in the increased PPI metabolism accompanying Ca^{2+} mobilization in stimulated cells. This is probably not the case as the evidence for distinct compartmentation of the PPIs is increasing. In platelets only 15 to 20% of the PIP₂ that is labeled with ^{32}P-Pᵢ in vitro is consumed during maximal stimulation with thrombin under conditions where PIP₂ resynthesis from PIP is blocked.[17,129] This suggests that 80% of the ^{32}P-PIP₂ is not available for the signal processing machinery, despite the fact this fraction is metabolically active; this phenomenon has been explained by treadmilling of putative PIP₂ polymers.[129] Incubation of erythrocytes with ^{32}P-orthophosphate labeled only the monoester phosphates, but not the diester phosphate, of PIP and PIP₂[177] which suggests that in these cells only the upper part of the reaction scheme (Figure 1, reactions 5 to 8) is operative, and only 25 to 60% of the total PPIs in erythrocytes appears to take part in these reactions.[178] Similarly, the monoester phosphates of the PPIs in platelets are labeled to a much greater extent than the diester phosphate during incubation of intact platelets with ^{32}P-Pᵢ.[125,129,179]

As discussed above, PI hydrolysis and PIP₂ hydrolysis have been suggested to be separate processes in stimulated platelets[124,125] and pancreas.[142] If this is the case, it is possible that the DG formed from PI is metabolically different from the DG formed from PIP₂, so that the PA formed subsequently is heterogenous. The ^{32}P/^{3}H-ratio of the PA that was formed during thrombin-treatment of platelets prelabeled with both ^{32}P-Pᵢ and ^{3}H-arachidonate[127] was not constant indicating a heterogeneity of PA. Figure 3A shows that thrombin also causes formation of PA in which the ^{32}P/^{3}H-ratio increased in platelets prelabeled with ^{32}P-Pᵢ and ^{3}H-glycerol. This increase is caused by formation of PA containing only ^{32}P and no ^{3}H in the first 30 sec of the thrombin-platelet interaction; subsequently ^{3}H-PA starts to be formed and accumulates at a constant rate after 60 sec (Figure 3B) and with a constant ^{32}P/^{3}H-ratio (Figure 3A). These changes could reflect formation of PA from DG with no ^{3}H in

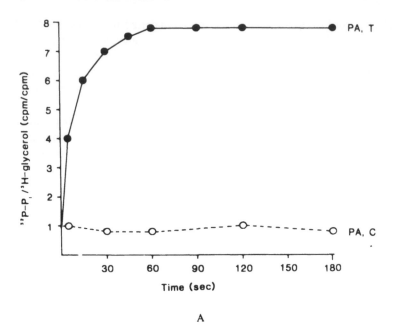

A

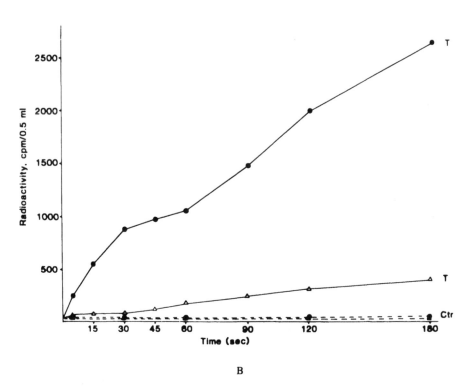

B

FIGURE 3.   Heterogeneity of PA formed in thrombin-stimulated platelets. Platelets were labeled in platelet-rich plasma and gel-filtered as described in Figure 2. Then the gel-filtered platelets were incubated for 1 hr (37°C) with 10 μCi/mℓ ^{3}H-glycerol. Such labeled platelets were incubated with a $^1/_{10}$ volume of thrombin (0.2 U/mℓ, solid lines) or 0.15 M NaCl (control, broken lines) and analyzed for ^{32}P- and ^{3}H-radioactivity in PA at the times (after addition of thrombin) shown. In (A) the ^{32}P/^{3}H-ratio in PA is shown, and in (B) the total radioactivities of ^{32}P ($\bullet$) and ^{3}H ($\triangle$) are given.

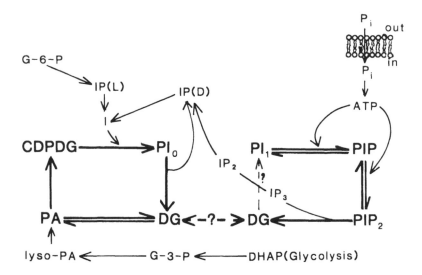

FIGURE 4. Possible separation of PPI metabolism from PI metabolism and entry of isotope by other metabolic pathways. The scheme illustrates the possibility which is discussed in the text, that PIP_2 is cleaved diesteratically and yields a DG which is converted (broken line) via PA to a small pool of PI (PI_1) which is metabolically interconverted to PIP and PIP_2. The major part of DG is, as hypothesized in the text, derived from a large pool of PI (PI_0) which is not metabolized to the PPIs; this DG is only reconverted via PA to PI_0. However, it is equally possible that DG from both PIP_2 and PI_0 are metabolically equal and "mix" with each other. The occurrence of heterogenous PA (Figure 3) could be explained by both mechanisms, but requires that in the *unstimulated* cell only the PI_0, and not PI_1, exchanges its glycerol moiety with added ^3H-glycerol. The scheme also illustrates the possibility that PA can be synthesized by two mechanisms: via the PPI/PI cycles and *de novo*, i.e., from dihydroxyacetone phosphate (DHAP) provided by glycolysis and via glycerol-3-phosphate (G-3-P) and lyso-PA. Interference in studies utilizing ^{32}P may also be experienced by a rapid entry of P_i (radioactive or not radioactive) which causes marked changes in the labeling of ATP, and therefore in the radioactivity of PIP, PIP_2 (upper right), and PA (not shown), that do not reflect changes in their mass. Similarly, in experiments using labeled inositol to prelabel the inositol lipids, inositol synthesized *de novo* from glucose-6-phosphate (G-6-P, upper left) or through exchange (not shown) may also cause radioactivity changes not related to changes in mass.

the first 30 sec, while PA formation from ^3H-DG began later with both types of DG originating through diesteratic hydrolysis of PI and or PIP_2, respectively.

However, heterogeneous PA could originate through other mechanisms. In stimulated neutrophils it has been claimed that only 2% of the PPIs is hydrolyzed by phospholipase C to DG and inositol phosphates, while the greater part of PI (12%) is thought to be converted to PA (with a specific ^{32}P radioactivity considerably lower than ATP) and to free inositol by phospholipase D.[93] Thus, some PA was made through DG phosphorylation while a larger portion was made directly from PI. Still another source of PA is through *de novo* synthesis, which has been suggested to increase during stimulation of secretion in platelets,[118,127] pancreas,[146] and salivary glands.[180]

The possible existence of different compartments of metabolically active PPIs, separate hydrolysis of PI and PIP_2 (thus we have a PI cycle *and* a PPI cycle) and the heterogeneity of DG and PA is shown schematically in Figure 4.

B. Different Metabolic Routes

The interpretation of results obtained in experiments where the metabolites in the PPI cycle have been labeled, is also difficult for reasons other than metabolic compartmentation. An unidentified phospholipid fraction that is labeled with radioactive phosphate, inositol, and arachidonate has been demonstrated in platelets and found to cochromatograph with PIP in several systems.[181] Interference may also come from nonenzymatic phosphorylation of PPIs by ATP, catalyzed by divalent cations to give PPI-pyrophosphates which may take place during tissue extraction.[182] Particularly when ^{32}P-labeling is used, the rapid uptake of P_i from the medium observed with stimulated platelets[126] and chemically transformed cells (which also show PPI response),[183] may rapidly alter the specific radioactivity of ATP and hence of PPIs and PA without changing their concentrations (Figure 4). When radioactive inositol is used to label the PPIs, the existence of a massive exchange of inositol in the inositol:phosphatidate transferase reaction (reaction no. 4, Figure 1; Figure 4) may cause marked changes in radioactivity that are not due to metabolism.[180,184,185] Interestingly, Mn^{2+} enhances this inositol exchange reaction and causes labeling of a specific pool of PI in brain that is agonist-insensitive.[184,185] In cells that synthesize inositol themselves, another source of error may be introduced when radioactive inositol is used to label the PPIs. IP (L -form) is formed from glucose-6-phosphate by the irreversible IP synthase reaction and converted by myoinositol-1-phosphatase to free inositol (meso-form) that is incorporated into PI (reaction no. 4 in Figure 1); the IP formed from PI or IP_2 is the D-form, which is also dephosphorylated by the phosphatase.[186] Thus, measurement of the turnover of the PPI cycle (or PI cycle) by determination of cycling of ^{3}H-inositol in cells prelabeled with this cyclohexanol could cause gross underestimations since the isotope would be diluted by newly synthesized inositol which is derived from glucose and thereby nonradioactive (Figure 4).

VI. CONCLUDING REMARKS

Even if the PPI cycle in all likelihood is necessary for stimulus-secretion coupling and contains steps that directly consume ATP, we are not presently in a position to estimate exactly the energy involved. Platelets are the most well-studied secretory cells in this respect, and *total* energy costs for complete dense granule secretion and complete α-granule + acid hydrolase secretion of 0.7 and 5.3 μmol ATP/10^{11} platelets have been estimated.[187,188] These values include, however, *all* energy-consuming processes that take place during the secretions, and which fraction of this is due to the PPI cycle, is not known. From the measured data of PI resynthesis[18] it was calculated above that 0.1 μmol ATP/10^{11} platelets was consumed per min by unidirectional (i.e. without action of phosphohydrolases) turnover of the PPI cycle, which is 50 times less than the energy consumed during acid hydrolase secretion which takes about 1 min.

Although this 50-fold difference in total ATP consumption and consumption during unidirectional turnover of the PPI cycle is based on one particular secretion in one particular cell, the example indicates that the energy consumption in the PPI cycle may only account for a very small fraction (2%) of the total energy consumption during secretion. However, the PPI cycle is not unidirectional as three "futile" cycles are operating, so that the actual energy consumption must be somewhat higher than 2% of total.

It is quite apparent from this literature review that our knowledge about the individual steps in the PPI cycle is very limited, and that a high degree of metabolic as well as subcellular compartmentation of the conversions apparently exists. This, and the extensive and rather uncritical use of prelabeling with different radioactive precursors, make the study of PPI metabolism particularly complicated. In addition, most of the metabolites in the PPI cycle are amphophilic molecules and present in micellar structures in the cell. It is therefore clear that much more knowledge about the particular steps in the PPI cycle, the compartmentation of the reactions, and general physical biochemistry of amphophilic molecules are required in future research in this area.

REFERENCES

1. **Michell, R. H., Kirk, C. J., Jones, L. M., Downes, C. P., and Creba, J. A.,** The stimulation of inositol metabolism that accompanies calcium mobilization in stimulated cells: defined characteristics and unanswered questions, *Phil. Trans. R. Soc. London Ser. B,* 296, 123, 1981.
2. **Laychock, S. G. and Putney, J. W.,** Roles of phospholipid metabolism in secretory cells, in *Cellular Regulation of Secretion and Release,* Conn, P. M., Eds., Academic Press, New York, 1982, 53.
3. **Berridge, M. J.,** Oncogenes, inositol lipids and cellular proliferation, *Biotechnology,* 4, 541, 1984.
4. **Fain, J. N.,** Activation of plasma membrane phosphatidylinositol turnover by hormones, *Vitam. Horm.,* 41, 117, 1984.
5. **Fisher, S. K., Van Rooijen, L. A. A., and Agranoff, B. W.,** Renewed interest in the polyphosphoinositides, *TIBS,* 9, 53, 1984.
6. **Hirasawa, K. and Nishizuka, Y.,** Phosphatidylinositol turnover in receptor mechanism and signal transduction, *Annu. Rev. Pharmacol. Toxicol.,* 25, 47, 1985.
7. **Streb, H., Irvine, R. F., Berridge, M. J., and Schulz, L.,** Release of Ca^{2+} from a nonmitochondrial intracellular store in pancreatic acinar cells by inositol-1,4,5-trisphosphate, *Nature (London),* 306, 67, 1983.
8. **Biden, T. J., Prentki, M., Irvine, R. F., Berridge, M. J., and Wollheim, C. B.,** Inositol 1,4,5-trisphosphate mobilizes intracellular Ca^{2+} from permeabilized insulin-secreting cells, *Biochem. J.,* 223, 467, 1984.
9. **Hirata, M., Suematso, E., Hashimoto, T., Hamachi, T., and Koga, T.,** Release of Ca^{2+} from a nonmitochondrial store site in peritoneal macrophages treated with saponin by inositol 1,4,5-trisphosphate, *Biochem. J.,* 223, 229, 1984.
10. **Gershengorn, M. C., Geras, E., Purello, V. S., and Rebecchi, M. J.,** Inositol trisphosphate mediates thyrotropin-releasing hormone mobilization of nonmitochondrial calcium in rat mammotropic pituitary cells, *J. Biol. Chem.,* 259, 10675, 1984.
11. **Lapetina, E. G., Watson, S. P., and Cuatrecasas, P.,** Myo-inositol 1,4,5-trisphosphate stimulates protein phosphorylation in saponin-permeabilized platelets, *Proc. Natl. Acad. Sci. U.S.A.,* 81, 7431, 1984.
12. **Adunyah, S. E. and Dean, D. L.,** Inositol trisphosphate-induced Ca^{2+} release from human platelet membranes, *Biochem. Biophys. Res. Commun.,* 128, 1274, 1985.
13. **Brass, L. F. and Joseph, S. K.,** A role for inositol trisphosphate in intracellular Ca^{2+} mobilization and granule secretion in platelets, *J. Biol. Chem.,* 260, 15172, 1985.
14. **Israels, S. J., Robinson, P., Docherty, J. C., and Gerrard, J. M.,** Activation of permeabilized platelets by inositol-1,4,5-trisphosphate, *Thrombosis Res.,* 40, 499, 1985.
15. **Authi, K. S., Evenden, B. J., and Crawford, N.,** Metabolic and functional consequences of introducing inositol 1,4,5-trisphosphate into saponin-permeabilized human platelets, *Biochem. J.,* 233, 707, 1986.
16. **Holmsen, H., Kaplan, K., and Dangelmaier, C. A.,** Differential energy requirements for platelet responses, *Biochem. J.,* 208, 9, 1982.
17. **Holmsen, H., Dangelmaier, C. A., and Rongved, S.,** Tight coupling of thrombin-induced acid hydrolase secretion and phosphatidate synthesis to receptor occupancy in human platelets, *Biochem. J.,* 222, 157, 1984.
18. **Tysnes, O. B., Verhoeven, A. J. M., and Holmsen, H.,** Phosphate turnover of phosphatidylinositol in resting and thrombin-stimulated platelets, *Biochim. Biophys. Acta,* 889, 182, 1986.
19. **Schacht, J.,** Purification of polyphosphoinositides by chromatography on immobilized neomycin, *J. Lipid Res.,* 19, 1063, 1978.
20. **Lodhi, S., Weiner, N. D., and Schacht, J.,** Interactions of neomycin with mononuclear films of polyphosphoinositides and other lipids, *Biochim. Biophys. Acta,* 557, 1, 1979.
21. **Palmer, F. B.,** Chromatography of acid phospholipids on immobilized neomycin, *J. Lipid Res.,* 22, 1296, 1981.
22. **Sastrasinh, M., Knauss, T. C., Weinberg, J. M., and Humes, H. D.,** Identification of the amino glycoside binding site in rat renal brush border membrane, *J. Pharmacol. Exp. Ther.,* 222, 350, 1982.
23. **Prentki, M., Deeney, J. T., Matschinsky, F. M., and Joseph, S. K.,** Neomycin: a specific drug to study the inositol-phospholipid signalling system? *FEBS Lett.,* 197, 285, 1986.
24. **Takenawa, T. and Nagai, Y.,** Purification of phosphatidylinositol-specific phospholipase C from rat liver, *J. Biol. Chem.,* 256, 6769, 1981.
25. **Hofmann, S. L. and Majerus, P. W.,** Identification and properties of two distinct phosphatidylinositol-specific phospholipase C enzymes from sheep seminal vesicular glands, *J. Biol. Chem.,* 257, 6461, 1982.
26. **Irvine, R. F. and Dawson, M. C.,** Phosphatidylinositol phosphodiesterase of rat brain: Ca^{2+}-dependency, pH optima and heterogeneity, *Biochem. J.,* 215, 431, 1983.
27. **Lenstra, R., Mauco, G., Chap, H., and Douste-Blazy, L.,** Studies on enzymes related to diacylglycerol production in activated platelets. I. Phosphatidylinositol-specific phospholipase C: further characterization using a simple method for detection of activity, *Biochim. Biophys. Acta,* 792, 199, 1984.

28. **Bennett, J. B., Cockcroft, S., Caswell, A. H., and Gomperts, B. D.,** Plasma-membrane location of phosphatidyl hydrolysis in rabbit neutrophils stimulated with formylmethionyl-leucylphenylalanine, *Biochem. J.,* 208, 801, 1982.

29. **Wilson, D. B., Bross, T. E., Hofmann, S. L., and Majerus, P. W,** Hydrolysis of polyphosphoinositides by purified seminal vesicle phospholipase C enzymes, *J. Biol. Chem.,* 259, 11718, 1984.

30. **Smith, C. D., Lane, B. C., Kusaka, I., Verghese, M. W., and Snyderman, R.,** Chemoattractant receptor-induced hydrolysis of phosphatidylinositol 4,5-bisphosphate in human polymorphonuclear membranes, *J. Biol. Chem.,* 260, 5875, 1985.

31. **Knight, D. E. and Scrutton, M. C.,** Cyclic nucleotides control a system which regulates Ca^{2+} sensitivity of platelet secretion, *Nature (London),* 309, 66, 1984.

32. **Haslam, R. J. and Davidson, M. M. L.,** GTP not cyclic GMP enhances secretion from permeabilized platelets, *Nature (London),* 313, 821, 1985.

33. **Haslam, R. J. and Davidson, M. M. L.,** Guanine nucleotides decrease the free Ca^{2+} required for the secretion of serotonin from permeabilized platelets. Evidence for a GTP-binding protein in platelet activation, *FEBS Lett.,* 174, 90, 1984.

34. **Cockcroft, S. and Gomperts, B. D.,** Role of guanine-binding protein in the activation of polyphosphoinositide phosphodiesterase, *Nature (London),* 314, 534, 1985.

35. **Wallace, M. A. and Fain, J. N.,** Guanosine-O-thiotriphosphate stimulates phospholipase C activity in plasma membranes of rat hepatocytes, *J. Biol. Chem.,* 260, 9527, 1985.

36. **Creutz, C. E., Dowling, L. G., Kyger, E. M., and Franson, R. C.,** Phosphatidylinositol-specific phospholipase C activity of chromaffin granule-binding proteins, *J. Biol. Chem.,* 260, 7171, 1985.

37. **Downes, C. P. and Michell, R. H.,** The polyphosphoinositide phosphodiesterase of erythrocyte membranes, *Biochem. J.,* 198, 133, 1981.

38. **Irvine, R. F., Letcher, A. J., and Dawson, R. M. C.,** Phosphatidylinositol-4,5-bisphosphate phosphodiesterase and phosphomonoesterase activities in rat brain, *Biochem. J.,* 218, 177, 1984.

39. **Sasaguri, T., Hirata, M., and Kuriyama, H.,** Dependence on Ca^{2+} of the activities of phosphatidylinositol 4,5-bisphosphate phosphodiesterase and inositol 1,4,5-trisphosphate phosphatase in smooth muscles of the porcine coronary artery, *Biochem. J.,* 231, 497, 1985.

40. **Van Rooijen, L. A. A., Sequin, E. B., and Aganoff, B. W.,** Phosphodiesteratic breakdown of endogenous polyphosphoinositides in nerve ending membranes, *Biochem. Biophys. Res. Commun.,* 112, 919, 1983.

41. **Loomis, C. R., Walsh, J. P., and Bell, R. M.,** sn-1,2-Diacylglycerolkinase of *E. coli.* Purification, reconstitution and partial amino- and carboxyterminal analysis, *J. Biol. Chem.,* 260, 4091, 1985.

42. **Kanoh, H. and Ohno, K.,** Partial purification and properties of diacylglycerol kinase from rat liver cytosol, *Arch. Biochem. Biophys.,* 209, 266, 1981.

43. **Kanoh, H., Kondoh, H., and Ono, T.,** Diacylglycerol kinase from pig brain. Purification and phospholipid dependencies, *J. Biol. Chem.,* 258, 1767, 1983.

44. **Sugimoto, Y., Whitman, M., Cantley, L. C., and Erikson, R. L.,** Evidence that the Rous sarcoma virus transforming product phosphorylates phosphatidylinositol and diacylglycerol, *Proc. Natl. Acad. Sci. U.S.A.,* 81, 2117, 1984.

45. **Kato, M., Homma, Y., Nagai, Y., and Takenawa, T.,** Epidermal growth factor stimulates diacylglycerol kinase in isolated plasma membrane vesicles from A431 cells, *Biochem. Biophys. Res. Commun.,* 129, 375, 1985.

46. **Call, F. L. and Williams, W. J.,** Phosphatidate phosphatase in human platelets, *J. Lab. Clin. Med.,* 82, 663, 1973.

47. **Holmsen, H. and Dangelmaier, C. A.,** Trifluoperazine stimulates thrombin-induced phosphatidate formation in platelets without inhibiting arachidonate liberation by phospholipase A_2: possible inhibition of phosphatidate phosphatase; *Biochem. Biophys. Res. Commun.,* submitted.

48. **Jamdar, S. C. and Osborne, L. J.,** Glycerolipid biosynthesis in rat adipose tissue. XI. Effect of polyamines on Mg^{2+} dependent phosphatidate phosphohydrolase, *Biochim. Biophys. Acta,* 752, 79, 1983.

49. **Call, F. L. and Williams, W. J.,** Biosynthesis of cytidine diphosphate diglyceride by human platelets, *J. Clin. Invest.,* 49, 392, 1970.

50. **Carter, J. R. and Kennedy, E. P.,** Enzymatic synthesis of cytidine diphosphate diglyceride, *J. Lipid Res.,* 7, 678, 1966.

51. **Agranoff, B. W., Bradley, R. M., and Brady, R. O.,** The enzymatic synthesis of inositol phosphatide, *J. Biol. Chem.,* 233, 1077, 1958.

52. **Yandrasitz, J. R. and Segal, S.,** The effect of $MnCl_2$ on the basal and acetylcholine-stimulated turnover of phosphatidylinositol in synaptosomes, *FEBS Lett.,* 108, 279, 1979.

53. **Schoepp, D. D.,** Manganese stimulates the incorporation of ^3H-inositol into a pool of phosphatidylinositol in brain that is not coupled to agonist-induced hydrolysis, *J. Neurochem.,* 45, 1481, 1985.

54. **Farkas, G., Enyedi, A., Sarkadi, B., Gardos, G., Nagy, Z., and Farago, A.,** Cyclic AMP-dependent protein kinase stimulates the phosphorylation of phosphatidylinositol to phosphatidylinositol-4-monophosphate in a plasma membrane preparation from pig granulocytes, *Biochem. Biophys. Res. Commun.*, 124, 871, 1984.

55. **Sarkadi, B., Enyedi, A., Farago, A., Meszaros, G., Kremmer, T., and Gardos, G.,** Cyclic AMP-dependent protein kinase stimulates the formation of polyphosphoinositides in lymphocyte plasma membrane, *FEBS Lett.*, 152, 195, 1983.

56. **Enyedi, A., Farago, A., Sarkadi, B., Szasz, I., and Gardos, G.,** Cyclic AMP-dependent protein kinase stimulates the formation of polyphosphoinositides in the plasma membranes of different blood cells, *FEBS Lett.*, 161, 158, 1983.

57. **Behar-Bannelier, M. and Murphy, R. K.,** An electrophoretic study of endogenous phosphorylation in vitro of the polypeptides of microsomal membrane fractions of mouse liver, *Biochem. J.*, 187, 147, 1980.

58. **Jergil, B. and Sundler, R.,** Phosphorylation of phosphatidylinositol in rat liver Golgi, *J. Biol. Chem.*, 258, 7968, 1983.

59. **Cockcroft, S., Taylor, J. A., and Judah, J. D.,** Subcellular localization of inositol lipid kinase in rat liver, *Biochim. Biophys. Acta*, 845, 163, 1985.

60. **Collins, C. A. and Wells, W. W.,** Identification of phosphatidylinositol kinase in rat liver lysosomal membranes, *J. Biol. Chem.*, 258, 2134, 1983.

61. **Smith, C. D. and Wells, W. W.,** Phosphorylation of rat liver nuclear envelopes. II. Characterization of in vitro lipid phosphorylation, *J. Biol. Chem.*, 258, 9368, 1983.

62. **Campbell, C. R., Fishman, J. B., and Fine, R. E.,** Coated vesicles contain a phosphatidylinositol kinase, *J. Biol. Chem.*, 260, 10848, 1985.

63. **Machicao, E. and Wieland, O. H.,** Evidence that the insulin receptor-associated protein kinase acts as phosphatidylinositol kinase, *FEBS Lett.*, 175, 113, 1984.

64. **Thompson, D. M., Cochet, C., Chambaz, E. M., and Gill, G. N.,** Separation and characterization of a phosphatidylinositol kinase activity that co-purifies with the epidermal growth factor receptor, *J. Biol. Chem.*, 260, 8824, 1985.

65. **Marche, P., Koutouzov, S., and Meyer, P.,** Metabolism of phosphoinositides in the rat erythrocyte membrane. A reappraisal of the effect of magnesium on the ^{32}P incorporation into polyphosphoinositides, *Biochim. Biophys. Acta*, 710, 332, 1982.

66. **Van Rooijen, L. A. A., Rossowska, M., and Bazan, N. G.,** Inhibition of phosphatidylinositol-4-phosphate kinase by its own product phosphatidylinositol-4,5-bisphosphate, *Biochem. Biophys. Res. Commun.*, 126, 150, 1985.

67. **Smith, C. D. and Wells, W. W.,** Phosphorylation of rat liver nuclear envelopes. I. Characterization of in vitro protein phosphorylation, *J. Biol. Chem.*, 258, 9368, 1983.

68. **Lundberg, G. A., Jergil, B., and Sundler, R.,** Subcellular localization and enzymatic properties of rat liver phosphatidylinositol-4-phosphate kinase, *Biochim. Biophys. Acta*, 846, 379, 1985.

69. **Van Dongen, C. J., Zwiers, H., and Gispen, W. H.,** Purification and partial characterization of the phosphatidylinsitol 4-phosphate kinase from rat brain, *Biochem. J.*, 223, 197, 1984.

70. **Van Dongen, C. J., Kok, J. W., Schrama, L. H., Oestriecher, B., and Gispen, W. H.,** Immunochemical characterization of phosphatidylinositol 4-phosphate kinase from rat brain, *Biochem. J.*, 233, 859, 1986.

71. **Van Dongen, C. J., Zwiers, H., De Graan, P. N. E., and Gispen, W. H.,** Modulation of the activity of purified phosphatidylinositol 4-phosphate kinase by phosphorylated and dephosphorylated B-50 protein, *Biochem. Biophys. Res. Commun.*, 128, 1219, 1985.

72. **Smith, C. D. and Wells, W. W.,** Characterization of a phosphatidylinositol 4-phosphate-specific phosphomonoesterase in rat liver nuclear envelopes, *Arch. Biochem. Biophys.*, 235, 529, 1984.

73. **Cooper, P. H. and Hawthorne, J. N.,** Phosphomonoesterase hydrolysis of polyphosphoinositides in rat kidney, *Biochem. J.*, 150, 537, 1975.

74. **Mack, S. E. and Palmer, F. B.,** Evidence for a specific phosphatidylinositol 4-phosphate phosphatase in human erythrocyte membranes, *J. Lipid Res.*, 25, 75, 1984.

75. **Lee, T. C. and Huggins, C. G.,** Triphosphoinositide phosphomonoesterase in rat kidney cortex. II. Purification and characterization, *Arch. Biochem. Biophys.*, 126, 214, 1968.

76. **Akhtar, R. A. and Abdel-Latif, A. A.,** Studies on the properties of triphosphoinositide phosphomonoesterase and phosphodiesterase of rabbit iris smooth muscle, *Biochim. Biophys. Acta*, 527, 159, 1978.

77. **Nijjar, M. S. and Hawthorne, J. N.,** Purification and properties of polyphosphoinositide phosphomonoesterase from rat brain, *Biochim. Biophys. Acta*, 480, 390, 1977.

78. **Roach, P. D. and Palmer, F. B.,** Human erythrocyte cytosol phosphatidyl-inositol bisphosphate phosphatase, *Biochim. Biphys. Acta*, 661, 323, 1981.

79. **Koutouzov, S. and Marche, P.,** The Mg^{2+}-activated phosphatidylinositol 4,5-bisphosphate-specific phosphomonoesterase of erythrocyte membrane, *FEBS Lett.*, 144, 156, 1982.

80. **Quist, E.,** Ca^{2+}-stimulated phospholipid phosphoesterase activities in rabbit erythrocyte membranes, *Arch. Biochem. Biophys.*, 236, 140, 1985.

81. **Raval, P. J. and Allan, D.,** Ca^{2+}-induced polyphosphoinositide breakdown due to phosphomonoesterase activity in chicken erythrocytes, *Biochem. J.,* 231, 179, 1985.
82. **Takenawa, T., Homma, Y., and Nagai, Y.,** Effect of calmodulin antagonists on lysosomal enzyme secretion and phospholipid metabolism in guinea-pig macrophages, *Biochem J.,* 208, 549, 1982.
83. **Holian, A. and Stuckle, R. N.,** Calcium regulation of phosphatidyl inositol turnover in macrophage activation by formyl peptides, *J. Cell. Physiol.,* 123, 39, 1985.
84. **Putney, J. W.,** Stimulus-permeability coupling: role of calcium in the receptor regulation of membrane permeability, *Pharmacol. Rev.,* 30, 209, 1978.
85. **Cockcroft, S., Bennett, J. P., and Gomperts, B. D.,** Stimulus-secretion coupling in rabbit neutrophils is not mediated by phosphatidyl breakdown, *Nature (London),* 288, 275, 1980.
86. **Cockcroft, S., Bennett, J. P., and Gomperts, B. D.,** The dependence on Ca^{2+} of phosphatidylinositol breakdown and enzyme secretion in rabbit neutrophils stimulated with formylmethionyl-leucylphenylalanine, *Biochem. J.,* 200, 501, 1981.
87. **Cockcroft, S.,** Ca^{2+}-dependent conversion of phosphatidylinositol to phosphatidate in neutrophils stimulated with fMet-Leu-Phe or ionophore A23187, *Biochim. Biophys. Acta,* 795, 37, 1984.
88. **Serhan, C. N., Broekman, M. J., Korchak, H. M., Smolen, J. E., Marcus, A. J., and Weissman, G.,** Changes in phosphatidylinositol and phosphatidic acid in stimulated human neutrophils. Relation to calcium mobilization, aggregation and superoxide radical generation, *Biochim. Biophys. Acta,* 762, 420, 1983.
89. **Takenawa, T., Homma, Y., and Nagai, Y.,** Role of Ca^{2+} in phosphatidylinositol response and arachidonic acid release in formylated tripeptide- or Ca^{2+} ionophore A23187-stimulated guinea pig neutrophils, *J. Immunol.,* 130, 2849, 1983.
90. **Hirayama, T. and Kato, I.,** A rapid stimulation of phosphatidylinositol metabolism in rabbit leukocytes by pseudomonal leukocidin, *FEBS Lett.,* 157, 46, 1983.
91. **Yano, K., Nakashima, S., and Nozawa, Y.,** Coupling of polyphosphoinositide breakdown with calcium efflux in formyl-methionyl-leucyl-phenylalanine-stimulated rabbit neutrophils, *FEBS Lett.,* 161, 296, 1983.
92. **Volpi, M., Yassin, R., Tao, W., Molski, T. F. P., Naccache, P. H., and Sha'afi, R. I.,** Leukotriene B_4 mobilizes calcium without the breakdown of polyphosphoinositides and the production of phosphatidic acid in rabbit neutrophils, *Proc. Natl. Acad. Sci. U.S.A.,* 81, 5966, 1984.
93. **Cockcroft, S., Barrowman, M., and Gomperts, B. D.,** Breakdown and synthesis of polyphosphoinositides in fMetLeuPhe-stimulated neutrophils, *FEBS Lett.,* 181, 259, 1985.
94. **Dougherty, R. W., Godfrey, P. P., Hoyle, P. C., Putney, J. W., and Freer, R. J.,** Secretagogue-induced phosphoinositide metabolism in human leukocytes, *Biochem. J.,* 222, 307, 1984.
95. **Ishizuka, Y., Imai, A., Nakashima, S., and Nozawa, Y.,** Evidence for *de novo* synthesis of phosphatidylinositol coupled with histamine release in activated mast cells, *Biochem. Biophys. Res. Commun.,* 111, 581, 1983.
96. **Ishizuka, Y., Imai, A., and Nozawa, Y.,** Polyphosphoinositide turnover in rat mast cells stimulated by antigen: rapid and preferential breakdown of phosphatidylinositol 4-phosphate (DPI), *Biochem. Biophys. Res. Commun.,* 123, 875, 1984.
97. **Nakamura, T. and Ui, M.,** Simultaneous inhibition of inositol phospholipid breakdown, arachidonic acid release, and histamine secretion in mast cells by islet-activating protein, pertussis toxin, *J. Biol. Chem.,* 260, 3584, 1985.
98. **Cohen, N. M., Schmidt, D. M., McGlennen, R. C., and Klein, W. L.,** Receptor-mediated increases in phosphatidylinositol turnover in neuron-like cell lines, *J. Neurochem.,* 40, 547, 1983.
99. **Harris, R. A., Fenner, D., and Leslie, S. W.,** Calcium uptake by isolated nerve endings: evidence for a rapid component mediated by the breakdown of phosphatidylinositol, *Life Sci.,* 32, 2661, 1983.
100. **Conn, P. J. and Sanders-Bush, E.,** Selective 5HT-2 antagonists inhibit serotonin stimulated phosphatidylinositol metabolism in cerebral cortex, *Neuropharmacology,* 23, 993, 1984.
101. **Bone, E. A., Fretten, P., Palmer, S., Kirk, C. J., and Michell, R. H.,** Rapid accumulation of inositol phosphates in isolated rat superior cervical sympathetic ganglia exposed to V_1-vasopressin and muscarinic cholenergic stimuli, *Biochem. J.,* 221, 803, 1984.
102. **Bone, E. A. and Michell, R. H.,** Accumulation of inositol phosphates in sympathic ganglia, *Biochem. J.,* 227, 263, 1985.
103. **Batty, I. R., Nahorski, S. R., and Irvine, R. F.,** Rapid formation of inositol 1,3,4,5-tetrakisphosphate following muscarinic receptor stimulation of rat cerebral cortical slices, *Biochem. J.,* 232, 211, 1985.
104. **Burgoyne, R. D.,** Mechanisms of secretion from adrenal chromaffin cells, *Biochim. Biophys. Acta,* 779, 201, 1984.
105. **Azila, N. and Hawthorne, J. N.,** Subcellular localization of phospholipid changes in response to muscarinic stimulation of perfused bovine adrenal medulla, *Biochem. J.,* 204, 291, 1982.
106. **Fisher, S. K., Holz, R. W. and Agranoff, B. W.,** Muscarinic receptors in chromaffine cell cultures mediate enhanced phospholipid labeling, but not catecholamine secretion, *J. Neurochem.,* 32, 491, 1981.

107. **Brown, J. H. and Masters, S. B.**, Does phosphoinositide hydrolysis mediate "inhibitory" as well as "excitatory" muscarinic responses?, *TIBS*, October, 417, 1984.

108. **Ohsako, S. and Deguchi, T.**, Phosphatidic acid mimicks the muscarinic action of acetylcholine in cultured bovine chromaffin cells, *FEBS Lett.*, 152, 62, 1983.

109. **Whitaker, M.**, Polyphosphoinositide hydrolysis is associated with exocytosis in adrenal medullary cells, *FEBS Lett.*, 189, 137, 1985.

110. **Farese, R. V., Larson, R. E., Sabir, M. A., and Gomez-Sanchez, G.**, Effects of angiotensin II and potassium and phospholipid metabolism in adrenal zona glomerulosa, *J. Biol. Chem.*, 256, 1093, 1981.

111. **Hunyady, L., Balla, T., Nagy, K., and Spat, A.**, Control of phosphatidylinositol turnover in adrenal glomerulosa cells, *Biochim. Biophys. Acta*, 713, 352, 1982.

112. **Hunyady, L., Balla, T., and Spat, A.**, Angiotensin II stimulates phosphatidylinositol turnover in adrenal glomerolusa cells by a calcium-dependent mechanism, *Biochim. Biophys. Acta*, 753, 133, 1983.

113. **Berridge, M. J., Downes, C. P., and Hanley, M. R.**, Lithium amplifies agonist-dependent phosphatidylinositol responses in brain and salivary glands, *Biochem. J.*, 206, 587, 1982.

114. **Balla, T., Enyedi, P., Hunyady, L., and Spat, A.**, Effects of lithium on angiotensin-stimulated phosphatidylinositol turnover and aldosterone production in adrenal glomerulosa cells: a possible causal relationship, *FEBS Lett.*, 171, 179, 1984.

115. **Enyedi, P., Buki, B., Mucsi, I., and Spat, A.**, Polyphosphoinositide metabolism in adrenal glomerulosa cells, *Mol. Cell. Endocrinol.*, 41, 105, 1985.

116. **Holmsen, H.**, Platelet metabolism and activation, *Semin. Hematol.*, 22, 219, 1985.

117. **Holmsen, H.**, Phospholipids: the phosphatidylinositol cycle, the polyphosphoinositide cycle, or phosphatidic acid?, in *Platelet Responses and Metabolism*, Vol. 3, Holmsen, H., Ed., CRC Press, Boca Raton, Fla., 1986, 121.

118. **De Chaffoy de Courcelles, D., Leysen, J. E., De Clerck, F., Van Belle H., and Janssen, P. A. J.**, Evidence that phospholipid turnover is the signal transducing system coupled to the serotonin-S_2 receptor sites, *J. Biol. Chem.*, 260, 7603, 1985.

119. **Billah, M. M., Lapetina, E. G., and Cuatrecasas, P.**, Phosphatidylinositol-specific phospholipase C of platelets: association with 1,2-diacylglycerol kinase and inhbition by cyclic AMP, *Biochem. Biophys. Res. Commun.*, 90, 92, 1979.

120. **Billah, M. M. and Lapetina, E. G.**, Degradation of phosphatidyl 4,5-bisphosphate is insensitive to Ca^{2+} mobilization in stimulated platelets *Biochem. Biophys. Res. Commun.*, 109, 1982.

121. **Watson, S. P., McConnell, R. T., and Lapetina, E. G.**, The rapid formation of inositol phosphates in human platelets by thrombin is inhibited by prostacyclin, *J. Biol. Chem.*, 259, 13199, 1984.

122. **Siess, W. and Binder, H.**, Thrombin induces the rapid formation of inositol bisphosphate and inositol trisphosphate in human platelets, *FEBS Lett.*, 180, 107, 1985.

123. **Siess, W.**, Evidence for the formation of inositol 4-monophosphate in stimulated human platelets, *FEBS Lett.*, 185, 151, 1985.

124. **Rittenhouse, S. E. and Sasson, J. P.**, Mass changes in myoinositol trisphosphate in human platelets stimulated by thrombin, *J. Biol. Chem.*, 260, 8657, 1985.

125. **Wilson, D. B., Neufeld, E. J., and Majerus, P. W.**, Phosphoinositide interconversions in thrombin-stimulated human platelets, *J. Biol. Chem.*, 260, 1046, 1985.

126. **Verhoeven, A. J. M., Tysnes, O. B., Horvli, O., Cook, C. A., and Holmsen, H.**, Thrombin-induced stimulation of phosphate uptake in human platelets, *J. Biol. Chem.*, 262, 7047, 1987.

127. **Holmsen, H., Dangelmaier, C. A., and Holmsen, H. K.**, Thrombin-induced platelet responses differ in requirement for receptor occupancy: evidence for tight coupling of occupancy and compartmentalized phosphatidic acid formation, *J. Biol. Chem.*, 265, 9393, 1981.

128. **DeChaffoy DeCourcelles, D., Roevens, P., and VanBelle, H.**, R 59 022, a diacylglycerol kinase inhibitor. Its effect on diacylglycerol and thrombin-induced C kinase activation in the intact platelet, *J. Biol. Chem.*, 260, 15762, 1985.

129. **Holmsen, H., Opstvedt Nilsen, A., and Rongved, S.**, Energy requirements for stimulus-response coupling, *Adv. Exp. Med. Biol.*, 192, 215, 1985.

130. **De Chaffoy de Courcelles, D., Roevens, P., and van Belle, H.**, 12-*O*-tetradecanoylphorbol 13-acetate stimulates inositol lipid phosphorylation in intact human platelets, *FEBS Lett.*, 173, 389, 1984.

131. **DeChaffoy De Courcelles, D., Roevens, P., and VanBelle, H.**, 1-Oleyl-2-acetyl glycerol (OAG) stimulates the formation of phosphatidylinositol 4-phosphate in intact human platelets, *Biochim. Biophys. Res. Commun.*, 123, 589, 1984.

132. **Watson, S. P. and Lapetina, E. G.**, 1,2-Diacylglycerol and phorbol ester inhibit agonist-induced formation of inositol phosphates in human platelets: possible implications for negative feedback regulation of inositol phospholipid hydrolysis, *Proc. Natl. Acad. Sci. U.S.A.*, 82, 2623, 1985.

133. **MacIntyre, D. E., McNicol, A., and Drummond, A. H.**, Tumour-promoting phorbol esters inhibit agonist-induced phosphatidate formation and Ca^{2+} flux in human platelets, *FEBS Lett.*, 180, 160, 1985.

134. **Tallanr, E. A. and Wallace, R. W.,** Calmodulin antagonists elevate the levels of ^{32}P-labeled polyphosphoinositides in human platelets, *Biochem. Biophys. Res. Commun.*, 131, 370, 1985.

135. **Opstvedt, A., Rongved, S., Aarsaether, N., Lillehaug, J., and Holmsen, H.,** Differential effects of chlorpromazine on secretion, protein phosphorylation and phosphoinositide metabolism in stimulated platelets, *Biochem J.*, 238, 159, 1986.

136. **Hokin, M. R. and Hokin, L. E.,** Enzyme secretion and the incorporation of ^{32}P into phospholipids of pancrease slices, *J. Biol. Chem.*, 203, 967, 1953.

137. **Hokin, L. E. and Hokin, M. R.,** Effects of acetylcholine on the turnover of phosphoryl units in the individual phospholipids of pancreas and brain cortex slices, *Biochim. Biophys. Acta*, 18, 102, 1955.

138. **Michell, R. H.,** Inositol phospholipids and cell surface receptor function, *Biochim. Biophys. Acta*, 415, 81, 1975.

139. **Halenda, S. P. and Rubin, R. P.,** Phospholipid turnover in isolated rat pancreatic acini, *Biochem. J.*, 208, 713, 1982.

140. **Orchard, J. L., Davis, J. S., Larson, R. E., and Farese, R. V.,** Effects of carbachol and pancreozymin (cholecystokinin-octapeptide) on polyphosphoinositide metabolism in the rat pancreas in vitro, *Biochem. J.*, 217, 281, 1984.

141. **Rubin, R. P., Godfrey, P. P., Chapman, D. A., and Putney, J. W.,** Secretagogue-induced formation of inositol phosphates in rat exocrine pancreas, *Biochem. J.*, 219, 655, 1984.

142. **Farese, R. V., Orchard, J. L., Larson, R. E., Sabir, M. A., and Davis, J. S.,** Phosphatidylinositol hydrolysis and phosphatidylinositol 4',5'-diphosphate hydrolysis are separable responses during secretagogue action in the pancreas, *Biochim. Biophys. Acta*, 846, 296, 1985.

143. **Dixon, J. F. and Hokin, L. E.,** The formation of inositol 1,2-cyclic phosphate on agonist stimulation of phosphoinositide breakdown in mouse pancreatic minilobules, *J. Biol. Chem.*, 260, 16068, 1985.

144. **Best, L. and Malaisse, W. J.,** Enhanced de novo synthesis of phosphatidic acid and phosphatidylinositol in pancreatic islets exposed to nutrient or neurotransmitter stimuli, *Arch. Biochem. Biophys.*, 234, 253, 1984.

145. **Best, L. and Malaisse, W. J.,** Phosphatidylinositol and phosphatidic acid metabolism in rat pancreatic islets in response to neurotransmitter and hormonal stimuli, *Biochim. Biophys. Acta*, 750, 157, 1983.

146. **Best, L. and Malaisse, W. J.,** Stimulation of phosphoinositide breakdown in rat pancreatic islets by glucose and carbamylcholine, *Biochem. Biophys. Res. Commun.*, 116, 9, 1983.

147. **Burgess, G. M., McKinney, J. S., Irvine, R. F., and Putney, J. W.,** Inositol 1,4,5-trisphosphate and inositol 1,3,4-trisphosphate formation in Ca^{2+}-mobilizing-hormone-activated cells, *Biochem. J.*, 232, 237, 1985.

148. **Michell, R. H. and Jones, L. M.,** Enhanced phosphatidylinositol labelling in rat parotid fragments exposed to a-adrenergic stimulation, *Biochem. J.*, 138, 47, 1974.

149. **Putney, J. W., Weiss, S. J., Van De Walle, C. M., and Haddas, R. A.,** Is phosphatidic acid a calcium ionophore under neurohormonal control?, *Nature (London)*, 284, 345, 1980.

150. **Jones, L. M. and Michell, R. H.,** The relationship of calcium to receptor-controlled phosphatidylinositol turnover, *Biochem. J.*, 148, 479, 1975.

151. **Weiss, S. J. and Putney, J. W.,** The relationship of phosphatidylinositol turnover to receptors and calcium-ion channels in rat parotid acinar cells, *Biochem. J.*, 194, 436, 1981.

152. **Weiss, S. J., McKinney, J. S., and Putney, J. W.,** Regulation of phosphatidate synthesis by secretagogues in parotid acinar cells, *Biochem. J.*, 204, 587, 1982.

153. **Weiss, S. J., McKinney, J. S., and Putney, J. W.,** Receptor-mediated net breakdown of phosphatidylinositol 4,5-bisphosphate in parotid acinar cells, *Biochem J.*, 206, 555, 1982.

154. **Downes, C. P. and Wusteman, M. M.,** Breakdown of polyphosphoinositides and not phosphatidylinositol accounts for muscarinic agonist-stimulated inositol phospholipid metabolism in rat parotid glands, *Biochem. J.*, 216, 633, 1983.

155. **Aub, D. and Putney, J. W.,** Metabolism of inositol phosphates in parotid cells: implications for the pathway of the phosphoinositide effect and for the possible messenger role of inositol trisphosphate, *Life Sci.*, 34, 1347, 1984.

156. **Irvine, R. F., Letcher, A. J., Lander, D. J., and Downes, C. P.,** Inositol trisphosphates in carbachol-stimulated rat parotid glands, *Biochem. J.*, 223, 237, 1984.

157. **Farese, R. V., Larson, R. E., and Sabir, M. A.,** Ca^{2+}-dependent and Ca^{2+}-independent mechanisms for phosphatidylinositol hydrolysis and ^{32}P-labeling during cholinergic stimulation of the rat submaxillary gland in vitro, *Arch. Biochem. Biophys.*, 219, 204, 1982.

158. **Fain, J. N. and Berridge, M. J.,** Relationship between hormonal activation of phosphatidylinositol hydrolysis, fluid secretion and calcium flux in the blowfly salivary gland, *Biochem. J.*, 178, 45, 1979.

159. **Berridge, M. J. and Fain, J. N.,** Inhibition of phosphatidylinositol synthesis and the inactivation of calcium entry after prolonged exposure of the blowfly salivary gland to 5-hydroxytryptamine, *Biochem. J.*, 178, 59, 1979.

160. **Fain, J. N. and Berridge, M. J.,** Relationship between phosphatidylinositol synthesis and recovery of 5-hydroxytryptamine-responsible Ca^{2+} flux in blowfly salivary glands, *Biochem. J.,* 180, 655, 1979.

161. **Litosch, I., Saito, Y., and Fain, J. N.,** 5-HT-stimulated arachidonic acid release from labeled phosphatidylinositol in blowfly salivary glands, *Am. J. Physiol.,* 243, C222, 1982.

162. **Berridge, M. J.,** Rapid accumulation of inositol trisphosphate reveals that agonists hydrolyse polyphosphoinositides instead of phosphatidylinositol, *Biochem. J.,* 212, 849, 1983.

163. **Litosch, I., Lee, H. S., and Fain, J. N.,** Phosphoinositide breakdown in blowfly salivary glands, *Am. J. Physiol.,* 246, C141, 1984.

164. **Sadler, K., Litosch, I., and Fain, J. N.,** Phosphoinositide synthesis and Ca^{2+} gating in blowfly salivary glands exposed to 5-hydroxytryptamine, *Biochem. J.,* 222, 327, 1984.

165. **Litosch, I. and Fain, J. N.,** 5-Methyltryptamine stimulates phospholipase C-mediated breakdown of exogenous phosphoinositides by blowfly salivary gland membranes, *J. Biol. Chem.,* 260, 16052, 1985.

166. **Sutton, C. A. and Martin, T. F. J.,** Thyrotropin-releasing hormone (TRH) selectively and rapidly stimulates phosphatidylinositol turnover in GH pituitary cells: a possible second step of TRH action, *Endocrinology,* 110, 1273, 1982.

167. **Schlegel, W., Roduit, C., and Zahnd, G.,** Thyrotropin releasing hormone stimulates metabolism of phosphatidyl inositol in GH$_3$ cells, *FEBS Lett.,* 134, 47, 1981.

168. **Rebbechi, M. J., Kolesnick, R. N., and Gershengorn, M. C.,** Thyrotropin-releasing hormone stimulates rapid loss of phosphatidylinositol and its conversion to 1,2-diacylglycerol and phosphatidic acid in rat mammotropic pituitary cells, *J. Biol. Chem.,* 258, 227, 1983.

169. **Canonico, P. L., Bonetti, A. C., Scapagnini, U., and MacLeod, R. M.,** Phosphatidylinositol cycle: a possible link between dopamine and thyrotropin-releasing hormone (TRH) in the control of lactin release in vitro, *Biogenic Amines* 1, 201, 1984.

170. **Raymond, V., Leung, P. C. K., Veilleux, R., Lefevre, G., and Labrie, F.,** LHRH rapidly stimulates phosphatidylinositol metabolism in enriched gonadotrops, *Mol. Cell. Endocrinol.,* 36, 157, 1984.

171. **Macphee, C. H. and Drummond, A. H.,** Thyrotropin-releasing hormone stimulates rapid breakdown of phosphatidylinositol 4,5-bisphosphate and phosphatidylinositol 4-phosphate in GH$_3$ pituitary tumor cells, *Mol. Pharmacol.,* 25, 193, 1984.

172. **Drummond, A. H., Bushfield, M., and Macphee, C. H.,** Thyrotropin-releasing hormone-stimulated (^{3}H)inositol metbolism in GH$_3$ pituitary tumor cells, *Mol. Pharmacol.,* 25, 201, 1984.

173. **Schlegel, W., Rodiut, C., and Zahnd, G. R.,** Polyphosphoinositide hydrolysis by phospholipase C is accelerated by thyrotropin releasing hormone (TRH) in clonal rat pituitary cells (GH$_3$ cells), *FEBS Lett.,* 168, 54, 1984.

174. **Lucas, D. O., Bajjalieh, S. M., Kowalchyk, J. A., and Martin, T. F. J.,** Direct stimulation by thyrotropin-releasing hormone (TRH) of polyphosphoinositide hydrolysis in GH$_3$ cell membrane by guanine nucleotide-modulated mechanism, *Biochem. Biophys. Res. Commun.,* 132, 721, 1985.

175. **Whitaker, M. and Aitchison, M.,** Calcium-dependent polyphosphoinositide hydrolysis is associated with exocytosis in vitro, *FEBS Lett.,* 182, 119, 1985.

176. **Godfrey, P. P. and Putney, J. W.,** Receptor-mediated metabolism of the phosphoinositides and phosphatidic acid in rat lacrimal acinar cells, *Biochem. J.,* 218, 187, 1984.

177. **Hawkins, P. T., Michell, R. H., and Kirk, C. J.,** Analysis of the metabolic turnover of the individual phosphate groups of phosphatidylinositol 4-phosphate and phosphatidylinositol 4,5-bisphosphate, *Biochem. J.,* 218, 785, 1984.

178. **Muller, E., Hegewald, H., Jaroszewicz, K., Cumme, G. A., Hoppe, H., and Frunder, H.,** Turnover of phosphomonoester groups and compartmentation of polyphosphoinositides in human erythrocytes, *Biochem. J.,* 235, 775, 1986.

179. **Holmsen, H. and Horvli, O.,** unpublished observations, 1986.

180. **Lupu, M. and Oron, Y.,** Neurotransmitter-caused increase in ^{3}H-inositol incorporation into phosphatidylinositol de novo synthesis *vs.* exchange, *FEBS Lett.,* 183, 133, 1983.

181. **Tysnes, O. B., Aarbakke, G. M., Verhoeven, A. J. M., and Holmsen, H.,** Thin-layer chromatography of polyphosphoinositides from platelet extracts: interference by an unknown phospholipid, *Thrombosis Res.,* 40, 329, 1985.

182. **Gumber, S. C. and Lowenstein, J. M.,** Nonenzymatic phosphorylation of polyphosphoinositides and phosphatidic acid is catalysed by bivalent metal ions, *Biochem. J.,* 235, 617, 1986.

183. **Kubota, Y., Inoue, H., and Yoshioka, T.,** Increased labelling of polyphosphoinositide in chemically transformed cells, *Biochim. Biophys. Acta,* 875, 1, 1986.

184. **Labarca, R., Janowsky, A., and Paul, S. M.,** Manganese stimulates incorporation of ^{3}H-inositol into an agonist-insensitive pool of phosphatidylinositol in brain membranes, *Biochem. Biophys. Res. Commun.,* 132, 540, 1985.

185. **Schoepp, D. D.,** Manganese stimulates the incorporation of ^{3}H-inositol into a pool of phosphatidylinositol in brain that is not coupled to agonist-induced hydrolysis, *J. Neurochem.,* 45, 1481, 1985.

186. **Parthasarathy, R. and Eisenberg, F.,** The inositol phospholipids: a stereochemical view of biological activity, *Biochem. J.,* 235, 313, 1986.
187. **Verhoeven, A. J. M., Mommersteeg, M. E., and Akkerman, J. W. N.,** Quantification of energy consumption in platelets during thrombin-induced aggregation and secretion. The coupling between platelet responses and the increment in energy consumption, *Biochem. J.,* 221, 777, 1984.
188. **Verhoeven, A. J. M., Mommersteeg, L. R., and Akkerman, J. W. N.,** Comparative studies on the energetics of platelet responses induced by different agonists, *Biochem. J.,* 236, 879, 1986.

Index

INDEX